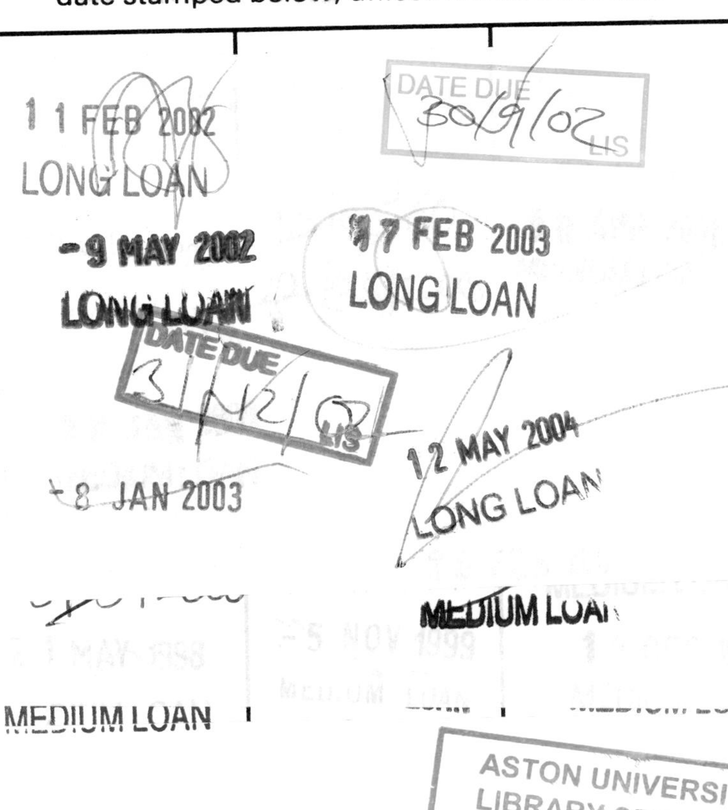
11 FEB 2002
LONG LOAN
DATE DUE
30/9/02
LIS
-9 MAY 2002
LONG LOAN
17 FEB 2003
LONG LOAN
DATE DUE
31/12/02
LIS
12 MAY 2004
LONG LOAN
-8 JAN 2003
MEDIUM LOAN
MEDIUM LOAN

ASTON UNIVERSITY
LIBRARY SERVICES
WITHDRAWN
FROM STOCK

Engineering catastrophes

Engineering catastrophes

Causes and effects of major accidents

JOHN LANCASTER

ABINGTON PUBLISHING

Woodhead Publishing Ltd in association with The Welding Institute
Cambridge England

Published by Abington Publishing, Abington Hall, Abington, Cambridge CB1 6AH, England

First published 1996. Abington Publishing
Reprinted in paperback 1997

British Library Cataloguing in Publication Data
A catalogue record for this book is available from the British Library.

ISBN 1 85573 342 0

Designed by Geoff Green (text) and The Colour Studio (cover).
Typeset by BookEns Ltd, Royston, Herts.
Printed by St Edmundsbury Press, Suffolk, England.

Contents

Preface

One of the privileges of the great, said Jean Giraudoux, is to witness catastrophes from a terrace. Giraudoux, who died in 1944, was a writer and diplomat, but he was not much of a prophet. Television has made nonsense of his words; today, it is the privilege of the multitude to witness catastrophes from an armchair. Come flood, fire, famine, storm, tempest, earthquake, volcanic eruptions or tidal waves, if the cameras can get there we will see it. No longer is it necessary to imagine suffering caused by disaster; it is there before us.

According to data collected by the United Nations, the cost of reconstruction following natural disasters, and the numbers of people killed or seriously affected by them, has risen during the last 30 years by an average of about 6% per year. This is to be compared with an annual population growth rate of 2%. Nobody knows why the effects of these catastrophes are increasing at such a rapid rate, but it is certain that they apply most severely to those countries that can least afford the cost. The years 1994–2004 have been declared 'the decade for natural disaster reduction'.

In the nineteenth century a new type of man-made catastrophe appeared. This was the start of the period of mass transportation, when railway accidents could, and sometimes did, result in large numbers of deaths. At the start of the century ships were relatively small so not many lives were lost when a single vessel foundered. Tonnages increased quite rapidly, however, and this period of growth culminated with the loss of over 1500 lives when the *Titanic* sank.

During the same period factories powered by steam engines grew in number and in size, as did the number of industrial accidents. One of the scandals of this period was the death rate due to boiler explosions. These peaked in about 1900 and then mercifully declined.

In the twentieth century we have seen the development of air transport, and the increasing size of aircraft has meant that a single loss could result in hundreds of deaths. Another big development has been in ferry transport, particularly roll-on, roll-off vehicle ferries. Here there have been some widely publicised catastrophes, notably the *Herald of Free Enterprise* and the *Estonia*. The worst ever shipping disaster was the sinking of a passenger ferry in the Philippines, with the loss of over 4000 lives.

More recently the search for oil and gas on the continental shelf has led to a completely new type of maritime activity. On mobile platforms in the North Sea it is a particularly dangerous one, comparable with deep-sea fishing so far as risk is concerned.

So where do we stand? Does the onward march of technology mean that more people are facing greater hazards, or is it otherwise?

It is the purpose of this book to try to answer the question by looking at the historical record of casualty and loss. War and pestilence have been excluded, and natural catastrophes are dealt with only in general terms, except, for reasons that will be set forth later, in the case of earthquakes. Most emphasis is on accidents to man-made objects and where possible the records that are kept on a worldwide basis will be used. It must be recognised, however, that such records refer almost entirely to ships, aircraft and so forth that are made in and operated by industrial countries.

I have tried as far as possible to avoid technical jargon. In particular the use of acronyms, except for a few old friends, has been avoided. This tiresome and unnecessary practice afflicts offshore technology more than most, such that some parts of Lord Cullen's report on the Piper Alpha disaster are incomprehensible, at least to the ordinary reader. Every effort has been made here to call a spade a spade.

Many of the disasters recorded in this book took place before the current (SI) system of units came into use. Where this is the case, the units quoted in the text are those given in the contemporary documents. A conversion table is provided in the Appendix in order that these older units may be transposed, should this be so required.

Acknowledgements

The variety of human artefacts that may suffer catastrophic failure, either due to operating error, to malfunction of the equipment itself, or to natural forces, is very great. To study such problems it is necessary to consult sources from a wide range of engineering disciplines. The present author has been fortunate enough to have access not only to a number of technical libraries but also to organisations, such as Lloyd's Register of Shipping in London and Marsh and McLellan in Chicago, whose statistics on failure rates (in shipping and hydrocarbon processing plant respectively) are unique. So thanks are due to the ever willing help of a number of librarians, and he would most particularly like to express appreciation for the assistance of TWI (formerly the Welding Institute) in Abington, Cambridge and the Institute of Marine Engineers in London for providing much valuable material.

Numbers of friends, colleagues and fellow technicians have given support and help, notably Ir A G de Koning of Shell, whose initial support started this project. Others who have contributed, either by providing material or support, or both, are Edmund Booth, N T Burgess, R J Cuninghame, Ted Piedenbrock and J R Spouge. And he is particularly grateful to his wife, Eileen, who was responsible for the typescript and for tackling unforeseen problems in obtaining photographs and copyright permissions.

Permission to reproduce copyright material is acknowledged as follows:

2.2/3 – The Ulster Folk & Transport Museum; 2.4 – Southampton City Heritage Services; 2.5 – Shelwing; 2.6/7 – RINA; 2.8–2.11 – Ministry of Justice, Oslo, Norway; 2.12/2.14 – HMSO; 2.16 – Grampian Police; 2.17/2.18 – RINA; 2.19–2.28 – HMSO; 3.4 – TWI; 3.10 – American Welding Society; 3.12 – Ministry of Justice, Oslo, Norway; 3.18–3.20 – British Petroleum; 3.25, 3.30/31 – TWI; 3.32–36 – Transport Research Laboratory and TWI; 4.6 – Cambridge University Press; 4.7 – American Welding Society; 4.8 – Ove Arup Partnership; 5.3 – TWI and Dr F Wallner; 5.9 – Cambridge University Library; 5.10/11 – HarperCollins; 5.12 – Oxford University Press; 5.13 – Cambridge University Library;

5.14 – Oxford University Press; 5.15 – British Aerospace; 5.17/18 – The Peninsular and Oriental Steam Navigation Company; 5.21 – British Petroleum; 5.24/25 – The Institute of Marine Engineers; 5.27 – British Petroleum.

CHAPTER 1

The historical record

Objectives and method

A primary objective in assembling statistics relative to major failures is to determine whether or not the proportional failure rate is increasing or decreasing, and by what amount. In many instances failure rates decrease with time, and although there is always a degree of scatter, it is possible to represent the data by a curve similar to that shown in Fig. 1.1. This figure relates to the case where the initial failure rate was 0.02 or 2% and the annual decrease or decrement in this rate is 5%; that is to say, at the end of one year it will have fallen to $0.02 \times 0.95 = 0.019$ or 1.9%. After ten years the failure rate falls to about 1.2%. Correspondingly the reliability, which is (1 − failure rate), starts at 0.980 and after 10 years has increased to 0.988.

Wherever practicable, the statistics quoted are for worldwide operations. In one or two cases, such as fatalities in manufacturing industry, figures for the United Kingdom have been used. In general these are representative of the USA and Western Europe, but are not necessarily true for, say, India or the developing nations of South-East Asia.

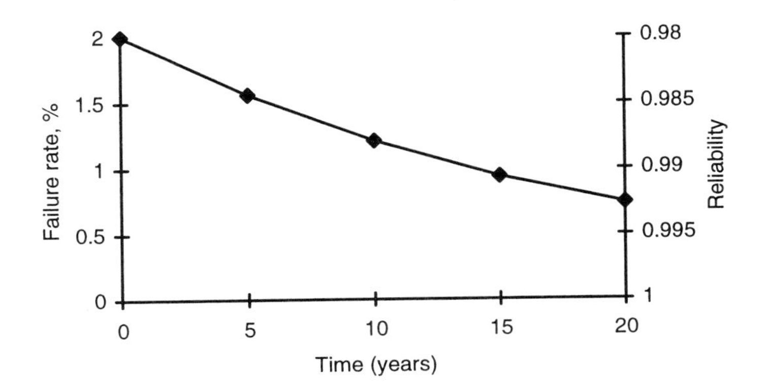

1.1 Exponential fall of failure rate. Initial rate 2%; annual decrement = 0.05.

Industrial fatality records include those due to catastrophes but are mainly concerned with accidents on a smaller scale. Nevertheless these and other fatality statistics are relevant because they can provide a measure as to what is, and what is not, an acceptable risk.

Fatalities in industry

Data falling under this heading have for the most part been taken from reports of the UK Health and Safety Commission. These cover manufacture, distribution, transport, financial services, public administration, education and other services; in other words, a very wide interpretation of 'industry'; and covering activities other than sport and leisure. This being so, it is worthwhile as a first step to consider fatality rates in those other areas.

Table 1.1 lists death rates in the UK due to natural causes and to a few types of accident. The population of the UK is about 50 million so these figures, except for the age groups, represent the annual death rate divided by 50. Road accidents are one of the very few cases where legislation has had a dramatic effect on fatality rates: prior to the compulsory use of seat belts the road accident death rate in the UK was about 7000 per year, while after the seat belt law had been passed it fell to 5000 per year.

Table 1.2 gives some risks associated with sport in the UK and the USA. The figures for the most hazardous sports are comparable with those for death by natural causes (primarily disease) in the (male) 35–44 age group. Both rates are similar to those for the most dangerous

Table 1.1 Annual fatality rate per million of the total population, other than where indicated, UK 1989 (after Warner[1])

Average	11 490
Men aged 55–64[a]	15 280
Women aged 55–64[a]	9 060
Men aged 35–44[a]	1 730
Women aged 35–44[a]	1 145
Boys aged 5–14[a]	225
Girls aged 5–14[a]	160
All accidents	240
Road accidents	98
Accidents in the home	86
Murder[b]	12
Murder by terrorists[c]	0.2

[a] Per million of age group.
[b] England and Wales 1990.
[c] England and Wales 1982–90.

Table 1.2 Annual fatality rate per million participants in various sports in Britain and the USA (after Warner[1])

Association football[a]	1.2
Motor sports[a]	27
Cave exploration, US, 1970–78	45
Rock climbing[a]	130
Glider flying, US 1970–78	400
Scuba diving, UK 1970–80	320
US 1970–78	420
Power boat racing, US 1970–78	800
Hang-gliding, UK 1977–79	1500
US 1978	400–1300
Sport parachuting US 1978	1900

[a] England and Wales 1986–90

Table 1.3 Annual fatality rate per million persons at risk in various UK industries (Warner[1])

	1974–78	1987–90
Clothing and footwear	5	0.9
Vehicle manufacture	15	12
Furniture	40	22
Brickworks, cement, etc.	65	60
Process plant operations	85	24
Shipbuilding	105	21
Agriculture (employees)	110	74
Construction	150	100
Railway staff	180	96
Coal mines	210	145
Offshore oil and gas[a]	1650	1250
Deep-sea fishing	2800	840

[a] Figures for worldwide operations are:[2] Offshore oil and gas, mobile units, 980 (1970–79), 1227 (1980–89).

occupations, deep-sea fishing and offshore mobile operations. Figures for occupational risk are given in Table 1.3.

The fall in the accident rate indicated in these data is reflected in those for all industry and services as recorded by the UK Health and Safety Commission[3] during the period 1961–91, and plotted in Fig. 1.2. In this and most other subsequent plots a regression analysis has been carried out on the natural logarithms of the failure rates, and the plotted trend curve is obtained from antilogarithms of the regression line.

Figure 1.2 includes the large number of fatalities that occurred when the Piper Alpha oil platform was destroyed. These are excluded in Fig. 1.3 and it will be evident that the effect on the trend curve is very small indeed. This would be expected because the population to which the

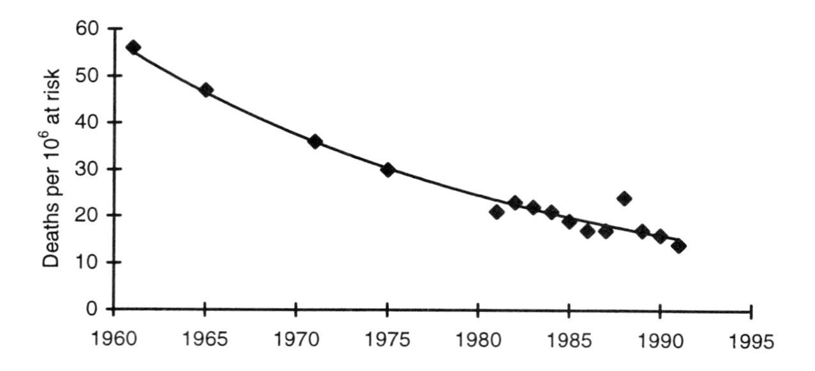

1.2 Annual rate of fatal injuries per million persons at risk in the UK 1961–91: employees in all industries. Annual decrement = 4.3%.

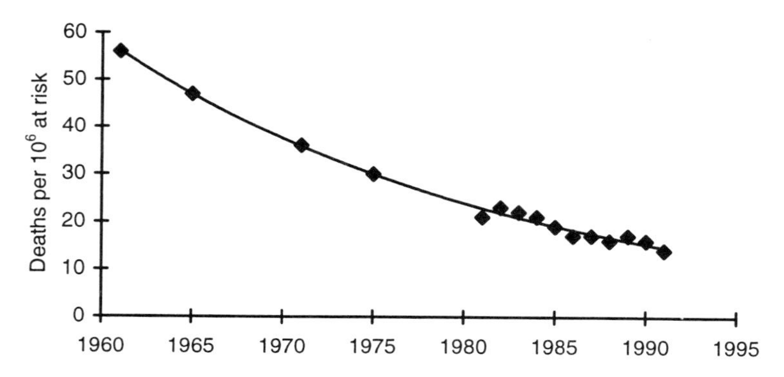

1.3 Annual rate of fatal injuries per 10^6 at risk in the UK: all industries, employees. Excludes Piper Alpha casualties. Annual decrement = 4.5%.

figure relates is high and the period under survey long. At the other extreme, where populations are smaller and the timescale shorter, this may not be the case. Figure 1.4 is a plot of the annual fatality rate in the United Kingdom energy and water supply industry for the period 1981–91. This plot includes the high value of 42.7 deaths per million in 1988 due to the Piper Alpha disaster, and the correlation is such as to indicate zero trend during the period. If the Piper Alpha figures are excluded (see Fig. 1.5) the plot then has a downward trend, but the correlation is not high enough to be confident that this is a real effect. On the other hand the general tendency in UK industry as a whole to show an annual reduction suggests that there is almost certainly a true fall.

The data shown so far are for employees. In the case of self-employed persons in industry (Fig. 1.6) no significant trend is to be seen for the

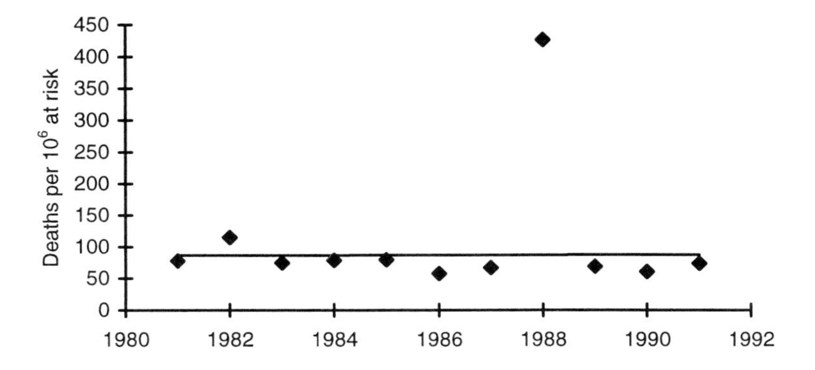

1.4 Annual rate of fatal injuries per million persons at risk in the UK 1981–91: energy and water supply industries, employees. Includes Piper Alpha fatalities. No significant change; mean = 87.1.

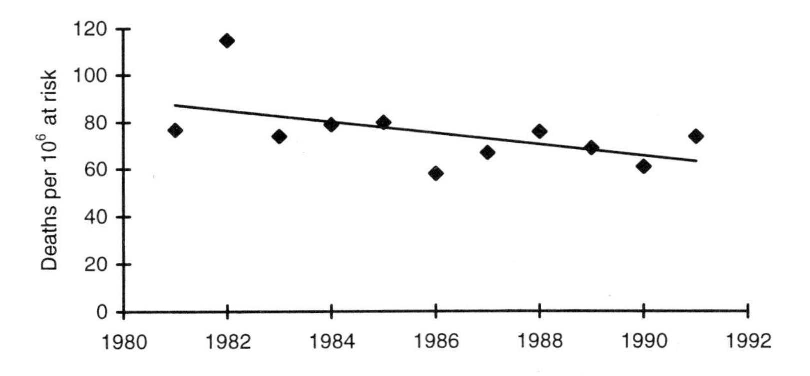

1.5 Annual rate of fatal injuries per million persons at risk in the UK 1981–91: energy and water supply industry, employees. Excludes Piper Alpha fatalities. Annual decrement = 3.0%.

period 1981–91, and the mean fatality rate is slightly higher than for industry as a whole during the same period. The timescale is too short for any firm conclusions to be drawn. It is almost certain that the casualty rate among self-employed people would have been higher in past decades, as in the case of employees (Fig. 1.2 and 1.3).

Other categories of workers covered by the UK Health and Safety Executive survey are those in agriculture and coastal fishing, construction and manufacturing industry. Data for employees in these three types of activity are plotted in Fig. 1.7, 1.8 and 1.9 respectively. They are consistent with expectations; not only is work in factories safer than in

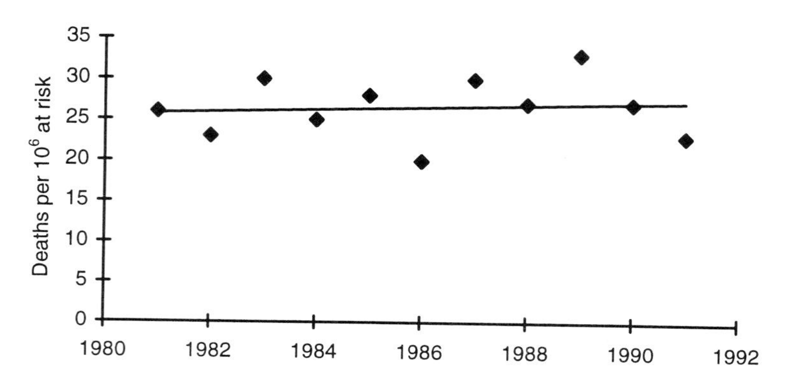

1.6 Annual rate of fatal injuries per million persons at risk in the UK 1981–91: self-employed in all industries. No significant change; mean = 25.9.

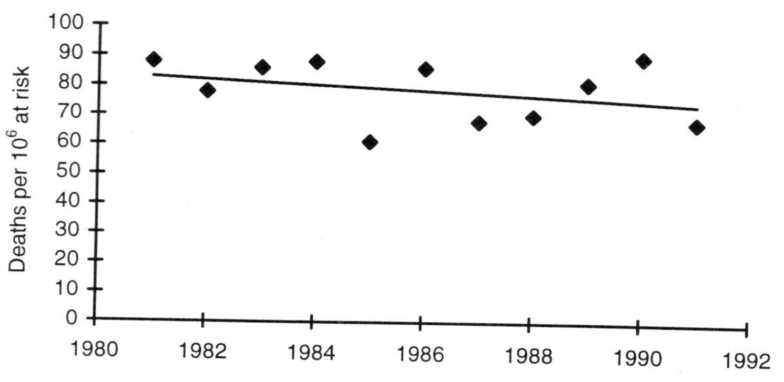

1.7 Annual rate of fatal injuries per million persons at risk in the UK 1981–91: agriculture and coastal fishing, employees. No significant change; mean = 77.8.

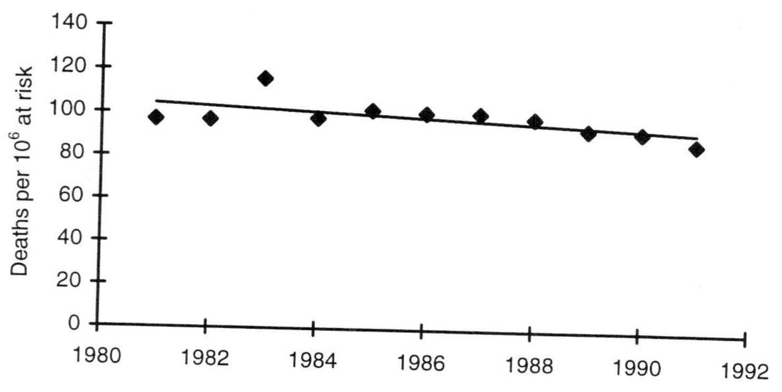

1.8 Annual rate of fatal injuries per million persons at risk in the UK 1981–91: construction, employees. Annual decrement = 1.2%.

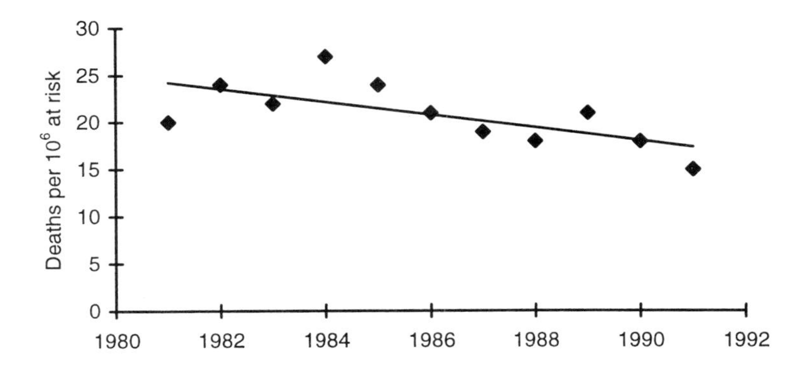

1.9 Annual rate of fatal injuries per million persons at risk in the UK 1981–91: manufacturing industry, employees. Annual decrement = 3.4%.

farms and fisheries, but there is an annual rate of improvement in safety in manufacture that is not apparent in agriculture during the 1981–91 period. Construction has the highest mean fatality rate, but does show a modest improvement during the decade. Factory work is by its nature the most controllable; safety regulations can be formulated and enforced more readily than in agriculture, and technological change is generally in the direction of lower risk. For example the automatic lathe, with rotary parts fully enclosed, replaces the more hazardous open lathe, robots take over from humans, and so forth. In farming, on the other hand, improved technology is slower to affect safety, and individual farmers are not under the watchful eye of a works safety committee. In this connection it is worth looking at the distribution of accidents by type. This analysis is available for the United Kingdom during the period 1986–92 and the figures expressed for a proportion of total fatalities are given in Table 1.4. The majority of accidents are either due to being struck by a moving object (a car, machinery, a falling spanner and the like) or falling from a height. Death due to fire or explosion is a relatively small risk under normal circumstances, but including the Piper Alpha fatalities would increase the percentage to about 12%. The underlying causes of these accidents, such as mechanical failure, natural hazard or human error, have not been investigated, but it would appear that most of them are the result of human error.

Comparing general and industrial risks

At the high end of the spectrum of fatality risk are certain sports, such as

Table 1.4 Fatal injury as a percentage of total fatalities, classified according to type of accident, UK 1986–92[3] (excludes Piper Alpha fatalities)

Type of accident	% of total
Collision (struck by falling object, moving vehicle, etc.)	37.2
Tripping, or falling from a height	26.5
Trapped by collapsing masonry or overturning machine	8.8
Drowning or asphyxiation	4.2
Exposure to harmful substance	2.4
Exposure to fire or explosion	4.5
Electrocution	6.4
Other	10.0
Total number of fatal accidents 1986–92, excluding Piper Alpha	2091

hang-gliding, and two occupations, offshore oil exploration drilling and deep-sea fishing, with an annual death risk in the region of 1 in 1000. It could be argued that comparing an occupation and a sport in this way is unreasonable because the hourly risk of a hazardous sport is much higher. However, annual risk is probably a more generally acceptable measure.

The most commonly accepted voluntary risk is that of using a motor car, where the fatality rate in 1989 was 1 in 10 000 per annum in the UK, and somewhat higher in other European countries and the USA. The figure for the construction industry is similar, while other industries are lower. If current trends continue, the overall risk in UK industry could well fall to 1 in 100 000 within a decade. Most occupations are now less hazardous than car driving, with some very much less.

Non-fatal accidents have not been considered here because reporting of such incidents is not compulsory; therefore the figures may or may not truly indicate trends.

Aircraft: losses and fatalities

In this and subsequent sections the statistics will include the rate of loss of hardware as well as the rate of loss of life.

Sources

There are two main sources for the material set out below. The first is the statistical survey which is published annually by the Boeing Corporation.[4] This survey covers civilian jet aircraft heavier than 60 000 lb (27 300 kg) gross weight worldwide, except for those manufactured in the former Soviet Union. It does not include accidents due to turbulence, boarding or disembarking, sabotage, military action or experimental flying. It *does* give details of the numbers of jet aircraft in the world fleet together with the annual numbers of departures, and losses are expressed as the number per million departures. This useful statistic enables a passenger to assess his or her chance of survival (which is, of course, very good; the fatality rate is 1 in 10 000 departures).

The other source is the Civil Aviation Authority's accident book, which is held at the CAA offices near Gatwick airport, England.[5] This document gives a detailed description of each and every accident to aircraft heavier than 5 700 kg maximum weight and is not restricted to jets. It is wider in scope than the Boeing survey but does not give total numbers of aircraft or departures. Both surveys allocate accidents to various categories: the CAA categories are much the more detailed and include sabotage, ground fire and destruction by fighter aircraft. It is comforting to know that this category accounts for only 3% of total losses.

Loss and fatality rates

Figure 1.10 shows the way in which jet aircraft numbers have increased since 1964, whilst Fig. 1.11 gives the number of annual departures during the same period. The passenger-carrying capacity of airplanes has also increased, so that while the number of craft has multiplied by ten, the number of passengers carried will have multiplied by a much larger number. Therefore calculations of the fatality rate based on aircraft numbers or numbers of departures are likely to be conservative. The relevant figures, as related to departures and aircraft numbers respectively, are shown in Fig. 1.12(a). There is inevitably a large variation from year to year in air accident fatalities; nevertheless there is a clearly defined

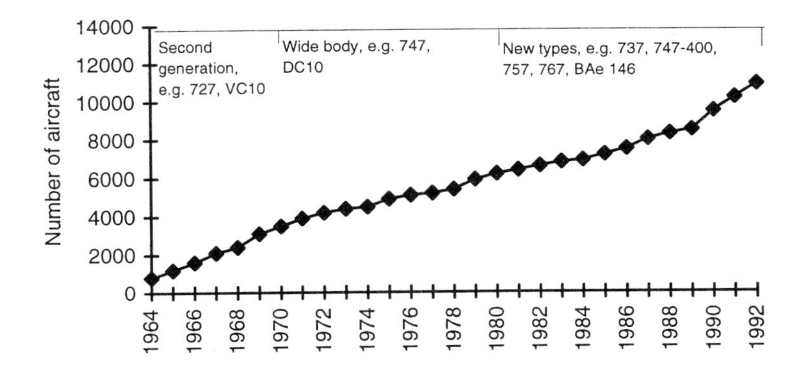

1.10 Jet aircraft in service worldwide 1964–92.

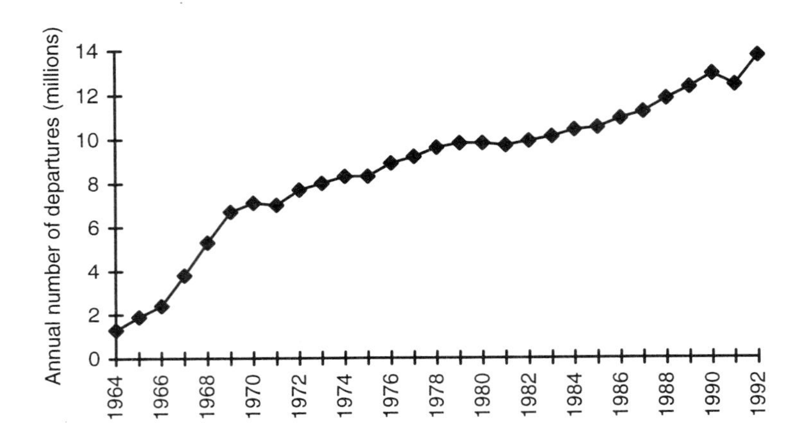

1.11 Annual number of departures of jet aircraft 1964–92.

downward trend with an annual decrement of the fatality rate which, for the period 1964–92 averages between 4 and 6%.

As would be expected, the figures for aircraft losses and accidents present a similar picture. Figure 1.13(a) gives the annual percentage of the world fleet of jet aircraft that were a total loss during the same period, whilst Fig. 1.13(b) shows equivalent rates for all accidents involving damage to the airplane. In the case of total losses the rate was about 0.6% at the start of the period, falling to somewhere between 0.1 and 0.2% at the end, corresponding to a mean annual decrement of about 4%.

The Boeing survey also includes statistics for accident rates (all accidents) per million departures of particular types of jet aircraft. These are plotted in Fig. 1.14, and show a typical 'bathtub' configuration with a high rate of early failures decreasing rapidly and then tending to increase again with time. Three points are worth noting. Firstly, the figures are for all accidents. Total losses (known in the aircraft industry as 'hull losses') range between 25 and 75% of all accidents, taken on an annual basis. Secondly, the peak annual loss rate is substantially higher than that for jet aircraft as a whole. Thirdly, the reasons for early failure may be quite different from those assumed in classical reliability theory, where it is supposed that a proportion of components are defective and therefore burn out or fail in some manner shortly after going into service. In aircraft, however, a high proportion of accidents are due to operational error, and the early failures may be due to the fact that crews lack experience with a new type of craft.

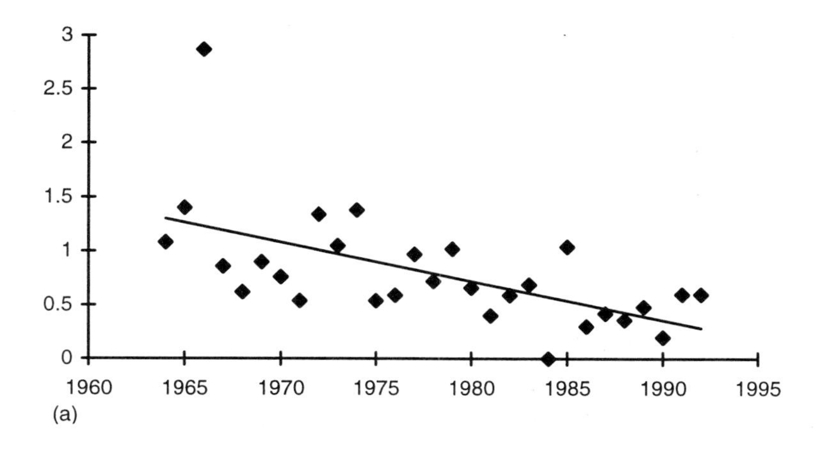

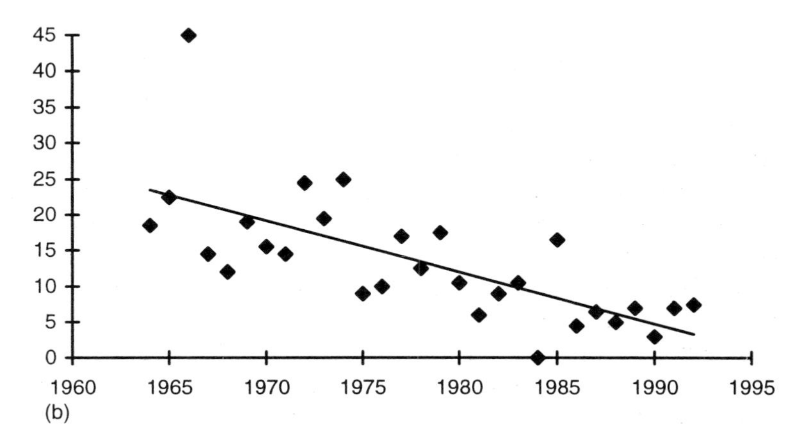

1.12 (a) Aircraft fatalities per 10 000 departures. Annual decrement = 4.1%. **(b)** Aircraft fatalities per 100 aircraft–years. Correlation coefficient of semi-ln plot = −0.7520; annual decrement = 5.3%.

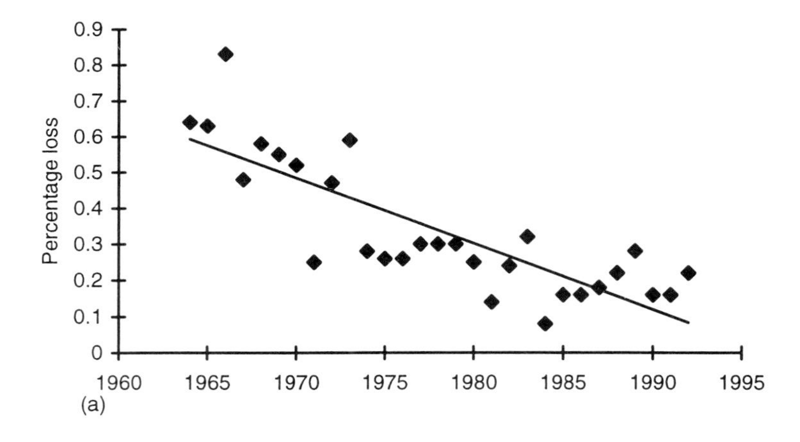

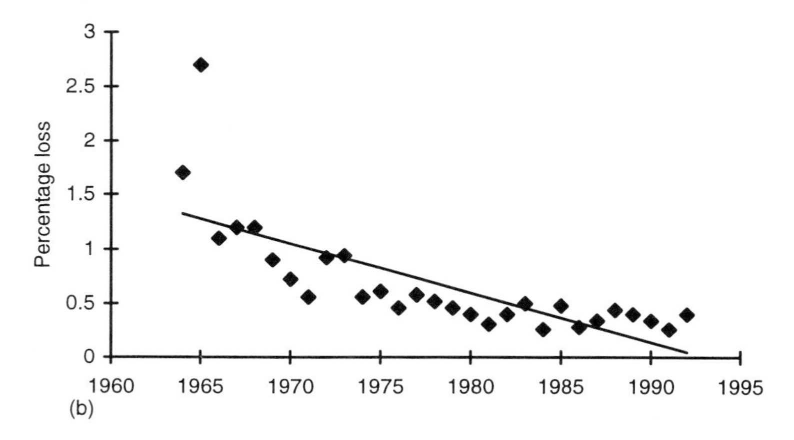

1.13 (a) Jet aircraft, total loss per year as a percentage of those in service. Annual decrement = 5.2%. (b) Jet aircraft, all accidents, per year as a percentage of those in service. Correlation coefficient of semi-ln plot = −0.8689; annual decrement = 5.9%.

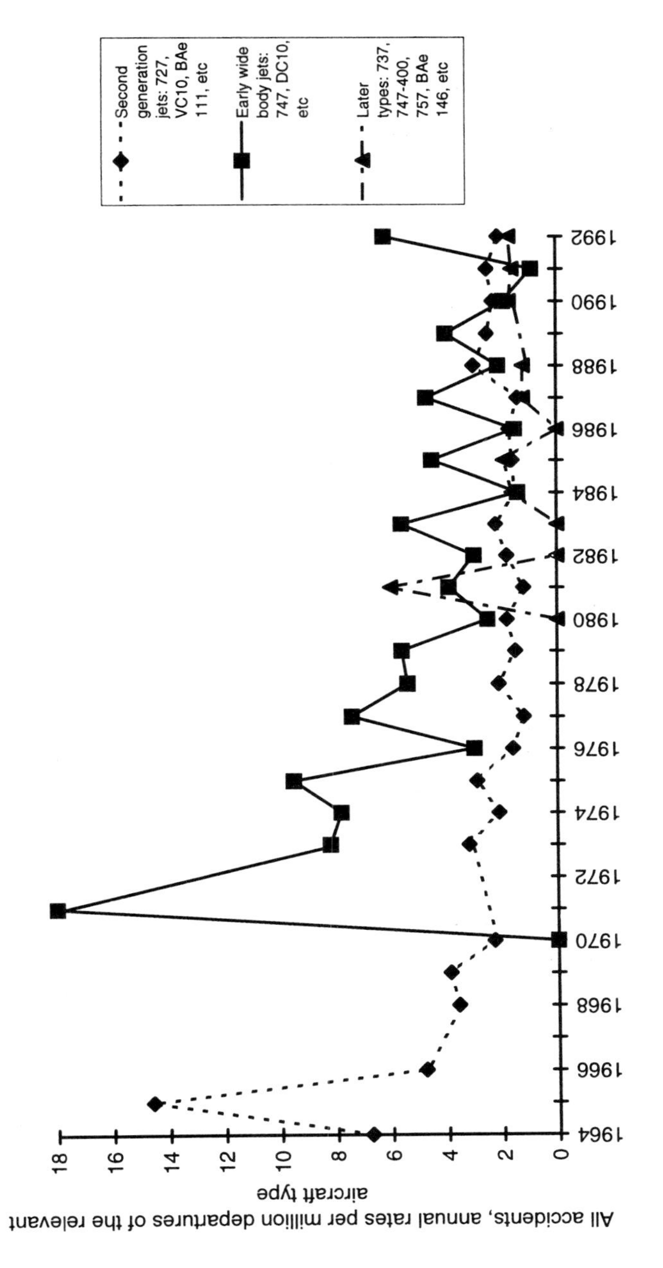

1.14 All accidents: successive generations of aircraft type.

Types and causes of loss

The CAA records adhere strictly to the descriptive type of category in designating the causes of aircraft losses. These are admirably detailed, and are listed in order of frequency in Table 1.5. This table confirms a well-established belief that the majority of losses occur on take-off or landing. The least probable cause is 'crew shot'. This category means exactly what it says: it includes the case of a disgruntled (and presumably deranged) ex-member of the airline staff who smuggled a pistol on board and proceeded to shoot the pilot, with disastrous consequences. Otherwise the list is in line with expectations, and indicates that operational error plays a major part in aircraft accidents.

This is corroborated by the Boeing survey, as shown in Fig. 1.15. The term 'Flightcrew' is not precisely defined, but is here assumed to mean 'accidents due to error by flightcrew'. The types of mechanical failure are analysed in Fig. 1.16, which indicates failure of power plant as the most important cause of losses of this type.

Table 1.5 Aircraft accidents categorised according to type, 1976–92

	Number	% of total
Overrunning/veering off runway	547	29.4
Collision with high ground	264	14.1
Collision with water	147	7.9
In-flight fire/smoke	117	6.3
Major power plant disruption	111	6.0
Failure of all power plants	103	5.6
Third party accident	97	5.2
Airframe failure	68	3.7
Ice/snow accumulation	49	2.6
Fuel exhaustion/mismanagement	46	2.5
Bird strike/ingestion	38	2.0
Shot down, ground fire or fighter	37	2.0
Aquaplaning	35	1.9
Mid-air collision	32	1.7
Electrical systems failure	30	1.6
Crew incapacitated	22	1.2
Sabotage: bomb on board	21	1.1
Door/window opening	20	1.1
Cargo shifting	11	0.6
Lightning strike	11	0.6
Tyre burst after retraction	8	0.4
Hail damage	5	0.3
Crew shot	4	0.2
Total	1696	100

Source: Civil Aviation Authority[5]

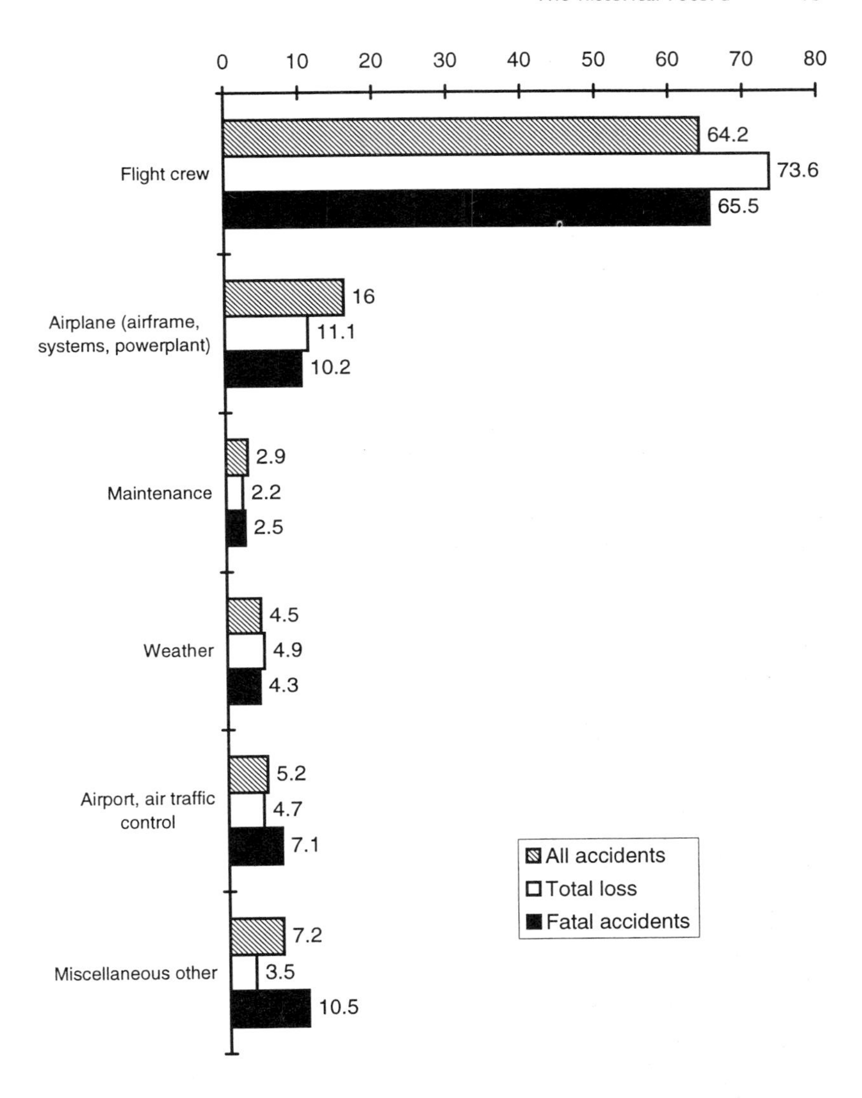

1.15 Cause of loss of jet aircraft 1959–92.[4] Percentage of losses by category: (a) all accidents; (b) total loss (hull loss); (c) accidents in which fatalities occurred.

The results of the two surveys are compared in Table 1.6. The various categories have been lumped together under three of the main underlying causes discussed earlier, namely operational error, mechanical failure and natural hazard (weather). Agreement is very good.

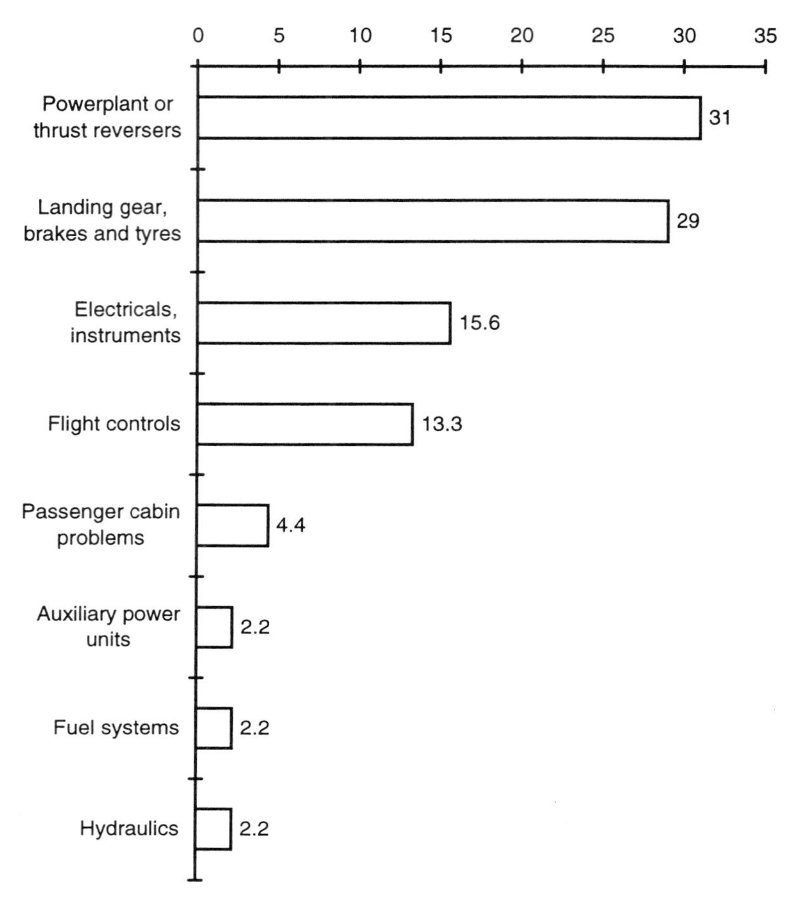

1.16 Type of mechanical failure leading to total loss of jet aircraft as a percentage of total failures. Boeing survey.[4]

Table 1.6 Aircraft accidents categorised according to cause; Boeing survey (jet aircraft) compared with Civil Aviation Authority log: all accidents as a percentage of total

	Boeing		CAA	
	No.	%	No.	%
Operational error	570	69	1036	61
Mechanical failure	55	19	340	20
Weather	37	5	100	6
Other	59	7	220	13
Total	721		1696	

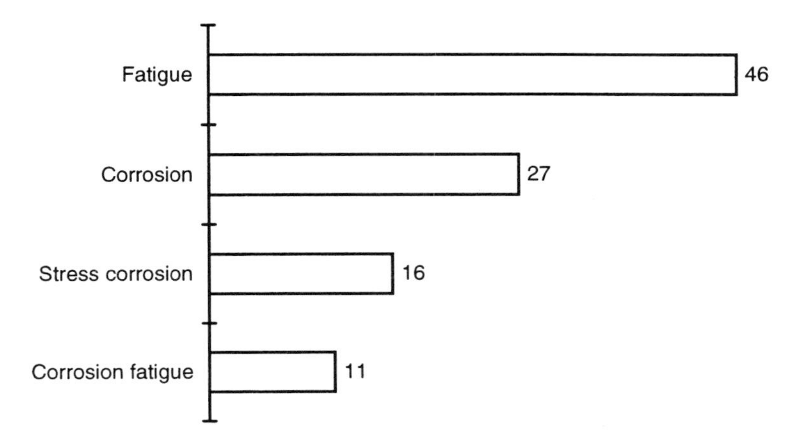

1.17 Percentage incidence of major failure mechanisms in aircraft.[6]

Failure mechanisms

Mechanisms of mechanical failure in aircraft have been investigated over a period of almost 50 years by the Royal Aerospace Establishment at Farnborough, England.[6] The failures in question include both those that precipitated an accident and those uncovered during routine inspection. The relative incidence of four major failure modes is shown in Fig. 1.17. Fatigue is indicated to be the most frequent of these modes, as in other areas of technology, but it is particularly important in aircraft, where numbers of fatal accidents have resulted from fatigue cracking. Corrosion is also a major problem in aircraft maintenance, although it rarely causes accidents. The aluminium structural alloys that are used in aircraft have poor corrosion resistance, and are protected by a layer of cladding in much the same way as galvanising protects steel. If this protection breaks down it is possible for large areas of metal to be corroded, resulting in costly repair work.

The other two failure modes, stress corrosion and corrosion fatigue, result, as in the case of fatigue, in the formation of cracks that may be difficult to find during inspection and which may result in a catastrophic failure. Therefore an important objective in aircraft design (and in the design of many other engineering structures) is to avoid features and exposures that may give rise to this type of failure.

Chemical, petroleum and petrochemical processing

In the nineteenth and early twentieth centuries the chemical and oil refining industries were separate entities, each with a distinct technology and method of operation. The chemical industry was coal-based, and in general conducted its operations inside factory buildings. The petroleum industry on the other hand was based on crude oil, the distillation of which was carried out in the open air, buildings being used mainly to house plant operators and office staff. In the second half of the twentieth century the situation changed. The bulk chemical industry went over to liquid or gaseous hydrocarbon feedstocks, and the plant adopted a similar layout to that of refineries. At the same time the production of organic chemicals from hydrocarbons – the petrochemical industry – became an important sector. As a result the chemical industry, with the exception of a few specialist areas, now shares a common technology with oil processing. This conglomerate can properly be called the hydrocarbon processing industry, and its activity is process plant operation.

As well as a shared technology, there are shared hazards. Hydrocarbons are inflammable and, when dispersed in air, may be explosive. Operating in the open reduces the risk of fire and explosion but does not eliminate it. Fortunately the risk to personnel is relatively low. The number of operators required even for a large plant is quite small, and in a properly designed unit those working in the control room are well protected. Table 1.3 shows that the annual fatality rate in the industry in recent years has (in the UK) fallen to 24 per million persons at risk.

Financial loss is another matter. Fire and explosion can severely damage large amounts of plant, and repair and replacement costs are high. In addition, there is loss of production over an extended period and there may be claims for incidental damage. The replacement costs alone appear to have increased about tenfold over the period 1963–92. During the same period the consumption of crude oil and natural gas increased somewhat less than three times. There is therefore an indication of increasing financial risk, which is a matter of serious concern.

Sources

A major source of quantitative information on losses in the hydrocarbon processing industry worldwide is the data bank held by Marsh and McLennan Protection Consultants of Chicago, originally compiled by W G Garrison and recently by D G Mahoney.[7] An analysis of 170 of the largest such losses that occurred during the period 1962–91 has been made by Mahoney.[8] Lees[9] has provided a comprehensive study of technical aspects of the problem. The causes of loss in the hydrocarbon processing, chemical and metal processing industries have been surveyed for the

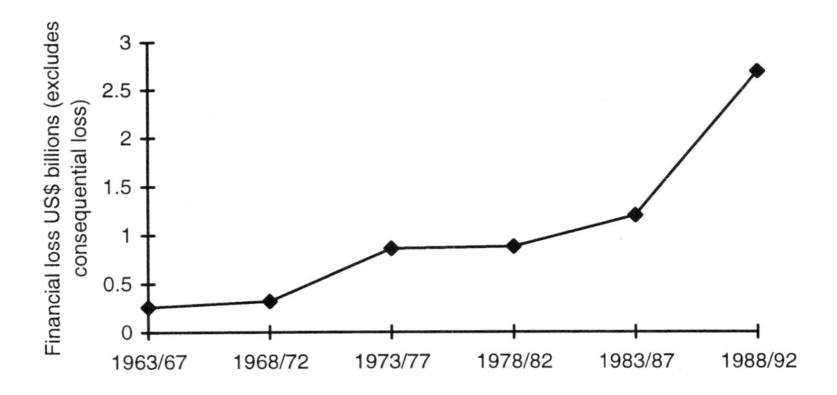

1.18 Capital cost of large process plant losses, worldwide 1963–92.[7]

European Community by Drogaris.[10] Material from this survey is summarised later.

The capital costs of the 100 largest process plant losses that occurred between 1963 and 1992 have been extracted from the 15th edition of *Large Property Damage Losses* and are shown in Fig. 1.18 as totals for successive five-year periods. These figures are replacement costs only; they do not include offshore or marine losses, or any consequential loss such as loss of production or third party claims. Most amounts above about $10 million are included; assuming that the distribution of capital values is normal (Gaussian), then about 90% of all losses are covered. There is no evidence of any slow-down in the incidence of loss; on the contrary, the rate would appear to be accelerating. The same applies, to a slightly lesser degree, to the number of incidents (Fig. 1.19).

The world production of crude oil and consumption of natural gas (in oil equivalent) are plotted in Fig. 1.20.[11] Figures for oil start in 1860, just after the first well was drilled at Oil Creek in Pennsylvania. They show a steady, exponential rise, little affected by the two world wars or the economic depression of the 1930s, until 1970. From 1970 to 1980 there was a modest reduction in the rate of increase, then in the early 1980s there was a fall followed by a slight recovery, such that production in the 1990s is about 5% higher than the previous peak.

Production of hydrocarbon processing plant followed a similar pattern. There was a construction boom in the post-1945 period for both refineries and petrochemical plant. The 1980s, however, were years of stagnation in process plant construction, and during this time once-prosperous contractors shrank to a fraction of their pre-1980 size. Refinery capacity worldwide fell by about 7% during this time. Production of natural gas, on

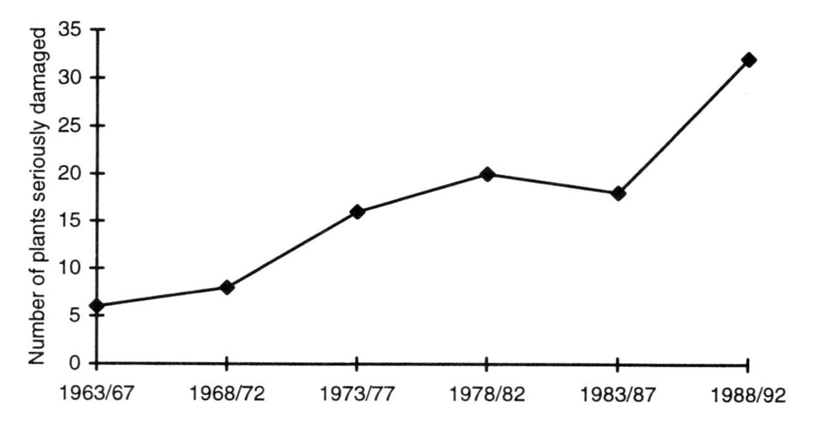

1.19 Number of large process plant losses, worldwide 1963–92.[7]

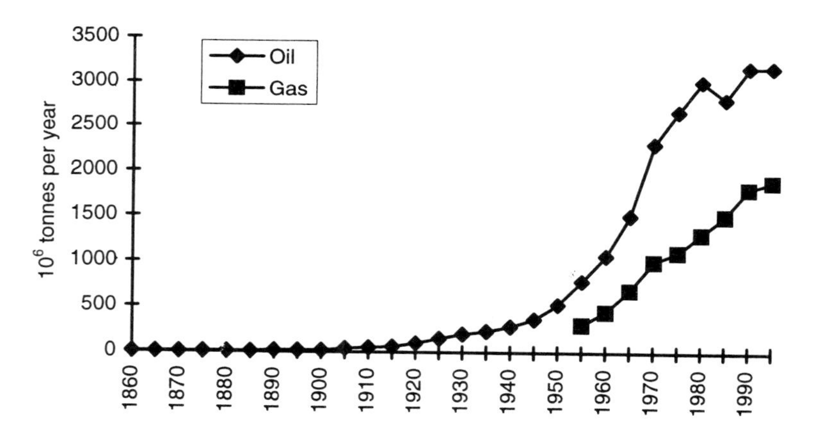

1.20 World crude oil production 1860–1993.[11] 1860 production = 70 000 tonnes. World natural gas consumption 1955–87: oil equivalent.

the other hand, has continued to increase and offshore exploration and drilling is still active.

A true assessment of process plant losses requires an estimate of the total value of such plant worldwide at the relevant time. There is a relationship between construction cost and plant capacity:

$$\text{Cost} = \text{constant} \times (\text{plant capacity})^n \quad [1.1]$$

where the value of n lies between 0.6 and 0.7. Thus, capital values will rise with overall process plant capacity, but at a slightly lower rate because there is a tendency for the size of individual units to increase. So very roughly the value of process plant worldwide could be expected to be proportional to the amount of hydrocarbon produced. The situation is complicated by the fact that a proportion of natural gas is used for heating and power generation. Also, most feedstocks for petrochemical process units have already been processed, either in natural gas treatment plants or (in the case of naphtha) in a refinery. Nevertheless, the sum of oil production plus the oil equivalent of gas production provides a reasonably good basis on which to see whether the true loss rate is improving or otherwise. In addition, by assigning a nominal value to the oil, it is possible to estimate what proportion of the industry's annual value is being destroyed. The price of oil is taken here to be $100 per tonne.

Figures for losses as a percentage of the nominal value of the oil and gas produced are plotted in Fig. 1.21. The line as drawn on this figure is the least-squares fit to the data as plotted; however, a better correlation is obtained with a semi-log plot, suggesting an exponential increase at a rate of about 5% annually.

Numbers of losses per billion tonnes of oil plus gas increase at a somewhat lower rate (Fig. 1.22). This suggests that the mean capital loss has increased during the period reviewed, and Fig. 1.23 shows that this is indeed the case.

Causes of accidents

The initial cause of major accidents to process plant were analysed both from the Marsh and McLennan (worldwide) data and from those collected from European countries in the EC survey. The results are plotted as a bar

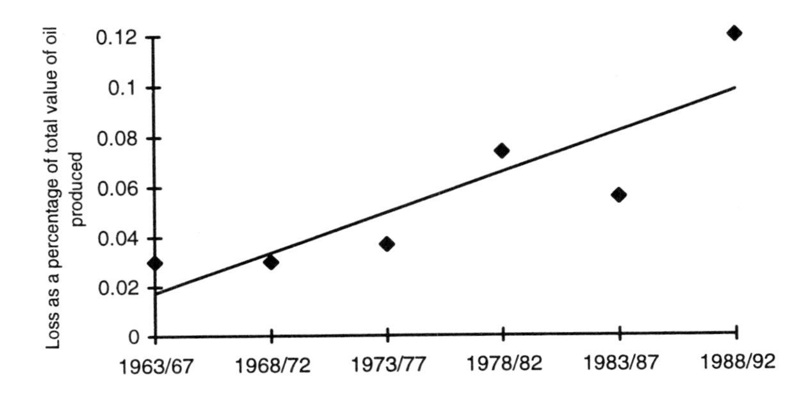

1.21 Capital value of large process plant losses as a percentage of the nominal value of hydrocarbon produced 1963–92.[7]

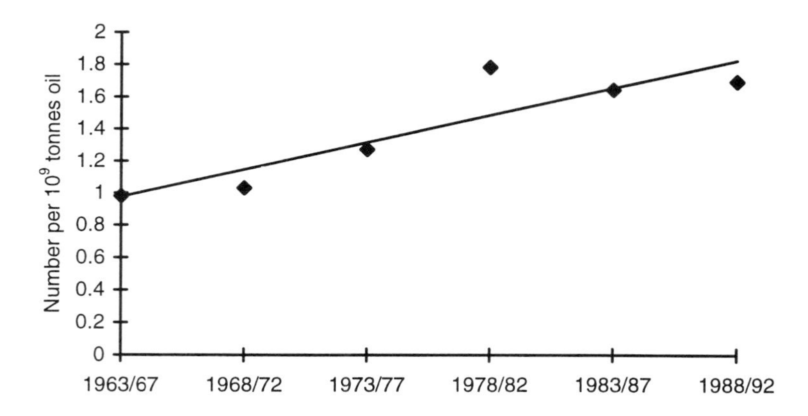

1.22 Number of large process plant losses per billion tonnes of oil and oil equivalent of gas produced 1963–92.[7]

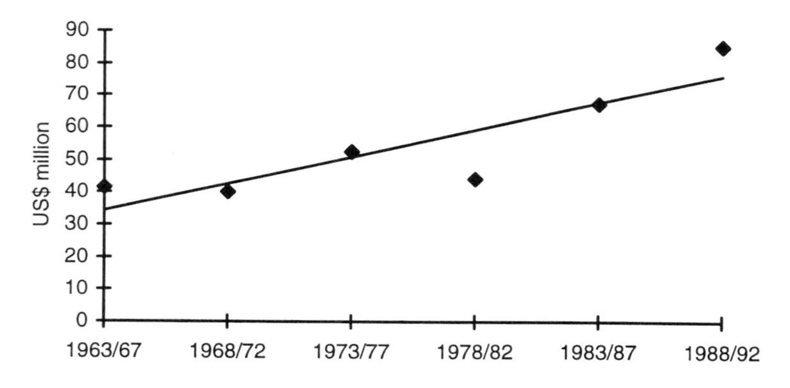

1.23 Mean capital value of major process plant losses 1963–92.

chart in Fig. 1.24. The agreement between these two sets of data is quite remarkable. Mechanical failure of equipment is the most prominent cause of major losses, followed by operational error. The largest single loss recorded up to the end of 1992 occurred in October 1989 at a high density polyethylene plant in Pasadena, Texas. A section of the plant was being isolated for maintenance, and this required that a pneumatically operated globe valve be fixed in the closed position and the air hoses disconnected. These operations were not carried out, and when the line was pressurised a large amount of ethylene and isobutane was released into the atmosphere. About one minute later the vapour cloud exploded with a force, according to seismographic data, of about 10 tons of TNT. Two high density polyethylene units and their associated equipment were destroyed, at a

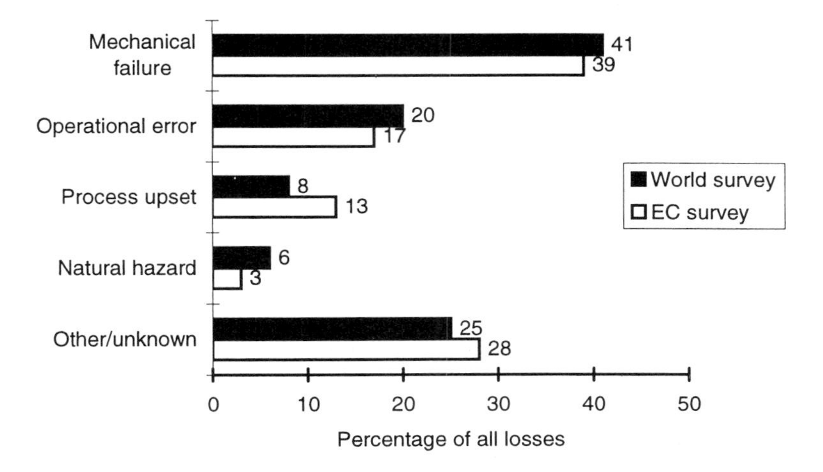

1.24 Cause of disruptive losses in hydrocarbon processing and chemical industries worldwide[8] and in Europe.[10]

total cost of US$715.5 million. The consequential loss amounted to almost the same figure.

In both surveys the type of equipment that failed was identified. For the worldwide survey all losses were so categorised, but in the European survey mechanical failures only were considered: these constituted 32% by number of all losses. The results are given in Tables 1.7 and 1.8. In both cases piping heads the lists. This is to be expected; in the first place there is more piping than any other category of equipment, and secondly it is exposed to virtually all the possible modes of degradation,

Table 1.7 Equipment in which failures leading to large losses in process plant occurred: worldwide survey, all losses[8]

Equipment	% of total number of losses	Average financial loss (US$ million)
Piping systems	31	41.9
Tanks	17	40.5
Reactors	13	28.9
Process drums	7	25.5
Marine vessels	6	32.0
Pumps/compressors	5	19.2
Heat exchangers	3	24.0
Towers	3	53.8
Heaters/boilers	1	28.6
Miscellaneous	9	34.7
Unknown	5	25.0

Table 1.8 Type of failure and equipment in which failure occurred: disruptive failures in European countries:[10] mechanical and corrosion failures only

	% of total number of losses
Mechanical failures	
Piping	25
Instruments and control systems	9
Valves	6
Machinery	6
Welds	5
Total, mechanical	82
Corrosion	
Internal	11
External	7
Total, corrosion	18

particularly corrosion and fatigue cracking. The failure modes themselves were not identified except that in the EC survey losses are divided into mechanical failure (82%) and corrosion (18%). However, non-disruptive failures have been so analysed, firstly for chemical process plant (Dupont in the USA)[12] and secondly for North Sea offshore processing equipment.[13] The results are shown as a bar chart in Fig. 1.25. Corrosion mechanisms are more frequent in non-disruptive failures, particularly for chemical plant. Fatigue predominates as a cause of mechanical breakdown in both surveys, as is the case in aircraft, and it is probably a major factor in mechanical failures that result in large process plant losses. Fatigue cracking is considered, quite rightly, as a most insidious mode of failure; it can be initiated by relatively small and seemingly unimportant changes in design, and it can progress unobserved until the pipe suddenly bursts or (as in the case of the *Alexander L Kielland* disaster) the structure collapses.

As indicated, corrosion is a less frequent cause of disruptive failure. It can happen, however, because of, for example, the use of incorrect material. During the construction of a crude oil distillation unit in a refinery it was found that there was a shortage of material for the piping which conveyed residual oil from the bottom of the atmospheric distillation tower to the vacuum tower. This line had been specified to be carbon steel to an ASTM (American Society for Testing and Materials) specification but to avoid delay the construction team made up the shortage with a piece of locally obtained pipe. Unfortunately this pipe had come from a rimming steel ingot, which meant that the outer surface was almost pure iron, but the bore was heavily contaminated with impurities.

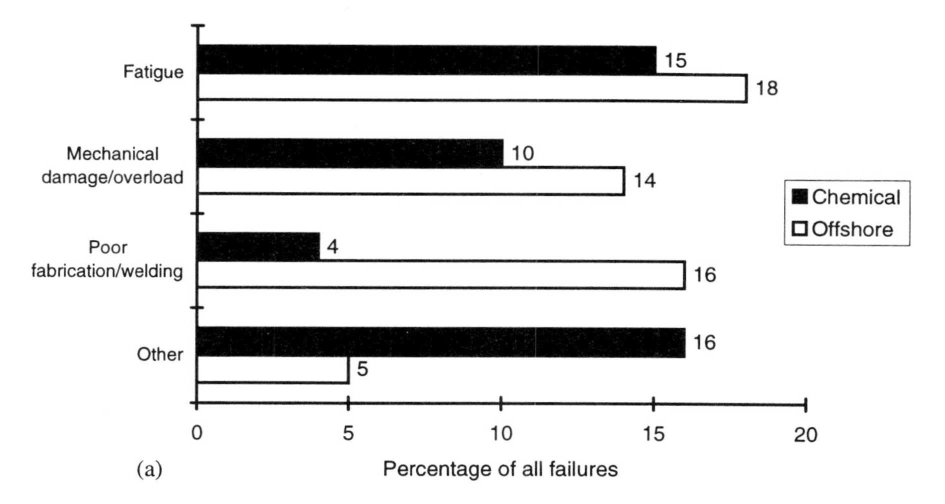

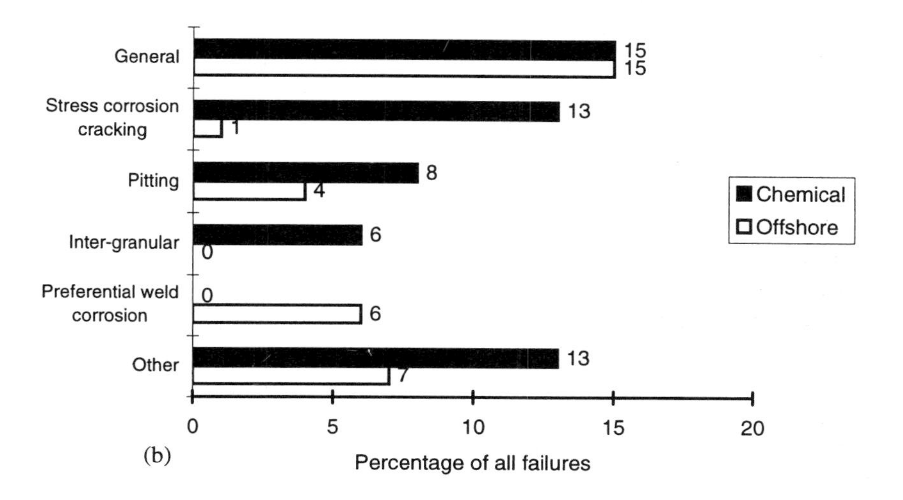

1.25 Type of non-disruptive failure in chemical[12] and offshore[13] process plant: (a) mechanical failures (total, chemical 45%, offshore 53%); (b) corrosion failures (total, chemical 55%, offshore 33%).

This impure material corroded rapidly and the pipe burst, causing a severe fire. It goes without saying that the provenance of all material used in such construction work should be strictly controlled.

Refineries are responsible for the greatest number of major losses (Table 1.9); naturally, there are more refineries than any other type of

Table 1.9 Types of process unit in which major losses occurred: worldwide survey,[8] 1962–91

Type of plant	% of total	Average capital loss US$ million
Refineries	40	44.6
Petrochemical plant	35	47.0
Terminals	11	28.1
Gas processing plant	6	55.5
Other	8	31.4

process unit. The highest average losses are for natural gas plants. Five of these losses occurred in very large plants in the Middle East; the size of the units combined with the hazards of handling light hydrocarbons would be expected to give this result. In earlier surveys the average pipeline loss was also high. This was largely due to the rupture in 1989 of a 28 inch hydrocarbon gas line carrying 13 500 tons per day from Siberia to petrochemical complexes in Western Russia. The operators observed a loss in pressure, and instead of shutting down, they tried to increase the pressure by starting up more pumps. As a result a very large vapour cloud formed and was ignited (it is thought) by sparks from electric locomotives operating in the vicinity. The resulting explosion was calculated to be equivalent to that due to 10 000 tonnes of TNT, and although it occurred in a rural location, caused damage estimated at US$150 million.

Vapour cloud explosions in general caused the highest average loss according to the 1992 analysis (Table 1.10) and when all types of plant are considered, were almost equal to fire as a cause of loss. One case, fortunately non-typical and unlikely to be repeated, concerned an ethylene plant that had recently been constructed and was in the commissioning stage. This plant, located in Europe, was adjacent to a large chemical complex which had contracted to purchase the entire output. During this early stage of operation the ethylene output, as would be expected, did not meet normal specifications for purity. A representative of the chemical company took a less charitable view, decided that he would no longer tolerate the situation, and closed a block valve in the transfer line to the factory. Almost immediately relief valves in the ethylene plant started to pop. Most of these valves, when open, fed into an overhead piping system that discharged into a stack which, for obvious reasons, was located remote from the furnaces. So the whole output of the unit rolled out of this stack as a vapour cloud and spread across the plant until it reached the furnaces and exploded. The situation was not improved by the fact that it was mid-winter. Fire engines from all around arrived to put out the fire, but the water from the fire hoses froze so that by the time the flames were out, the

Table 1.10 Type of incident causing major losses in process plant: worldwide survey,[8] 1962–91

(a) Type of loss

	Number	%	Average loss (US$ million)
Fire	62	36	36.1
Vapour cloud explosions	59	35	59.6
Internal and other explosions	43	25	33.6
Other	6	4	24.7
Total	170	100	43.2

(b) Type of loss by type of plant: percentage of total

	Fire	Vapour cloud explosion	Internal explosion	Other
Refineries	48	31	15	6
Petrochemical plant	17	37	46	0
Terminals	44	28	22	6
Gas processing plant	40	60	0	0
Other	50	36	7	7

towers and other equipment were encased in ice. Two bodies were later recovered when the ice melted; operators in the control room survived but some office workers in nearby buildings were badly cut by flying glass. The fate of the zealous inspector who closed the valve is not known.

Reformer tube failures

It is exceptional to find statistical reports on individual items of process plant. The exception is that of steam–methane reformer furnaces, on which a number of reports appeared in the early 1970s.

Steam–methane reforming is a means of producing hydrogen (for ammonia synthesis for example) by reacting steam with methane. The products of this reaction are carbon dioxide, carbon monoxide and hydrogen. The process is carried out in catalyst-packed tubes assembled in a furnace, as shown diagrammatically in Fig. 1.26. The operating conditions are severe: in a typical example the pressure is 500 psi with temperatures up to 950 °C. The material most commonly used for the tubes in early installations was centrifugally cast 25Cr20Ni alloy with 0.4% carbon, and the tubes were designed to have a life of 100 000 hours. The design stress was established by carrying out elevated temperature

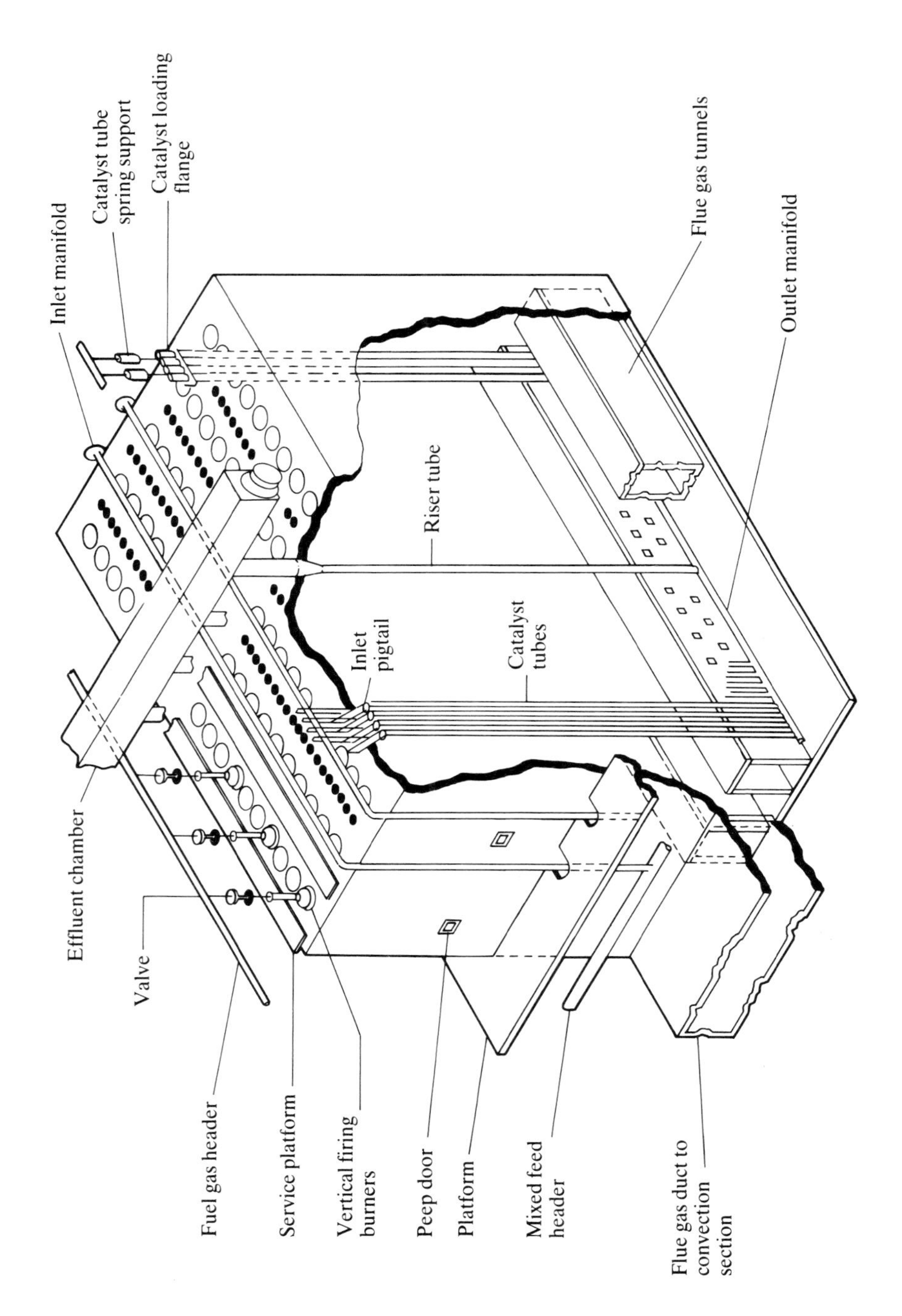

1.26 Top-fired steam–methane reformer furnace.

tests at constant loads for shorter times and extrapolating the results; there was thus some uncertainty about the true rupture stress. In addition, there is an inherent scatter of creep-rupture test results. And finally, it was found that a number of catalyst tubes were failing by creep-rupture long before their due term. Therefore a number of surveys of reformer tube failures were carried out, one by the present author[14] for top fired furnaces in Europe, and two in the USA, the first for side and bottom-fired[15] and the second for top-fired plants.[16] In the event it turned out that there was no significant difference in the behaviour of different designs of furnace.

The Weibull distribution

In all cases the data was analysed in accordance with the Weibull model.[17] This model relates the cumulative proportion of failures to time:

$$\frac{N_f}{(N_o+1)} = 1 - \exp\left[-\left(\frac{t}{t_m}\right)^b\right] \quad [1.2]$$

where t is elapsed time b is the Weibull exponent, t_m is the time corresponding to 63.2% failures, N_f is the number of failures at time t and N_o the original number of items. In reformer furnaces N_o is large and N_f is usually small, so

$$\frac{N_f}{(N_o+1)} \approx \frac{N_f}{N_o} = P \quad [1.3]$$

where P is the failure probability at time t. Equation 1.2 may be rearranged to give

$$\left(\frac{t}{t_m}\right)^b = \ln\left[\frac{1}{(1-P)}\right] \quad [1.4]$$

Putting $\ln [1/(1 - P] = p$ and taking logarithms

$$b\,(\log t - \log t_m) = \log p \quad [1.5]$$

When P, the cumulative proportion of failures, is small,

$$\ln [1/(1 - P)] \approx \ln (1 + P) \approx P \quad [1.6]$$

and this holds good for values of P below about 0.1.

It is one of the virtues of the Weibull model that in many practical cases, a logarithmic plot of cumulative proportion of failure against time yields a straight line. Sometimes early failures, of the type that produce the bathtub curve, may distort the picture, but these may justifiably be eliminated. It is then possible to extrapolate the plot to predict the time for any specified proportion of failures.

If we put $t = t_m$ in equation 1.2 then

$$N_f / (N_o + 1) = 1 - e^{-1} = 0.632 \qquad [1.7]$$

whence the definition of t_m.

Reformer tube data

A Weibull plot of data from the European survey is shown in Fig. 1.27. A total of 37 tube failures was reported, and of these 30 occurred in three of the plants and 7 in the remaining seven. The three plants with the

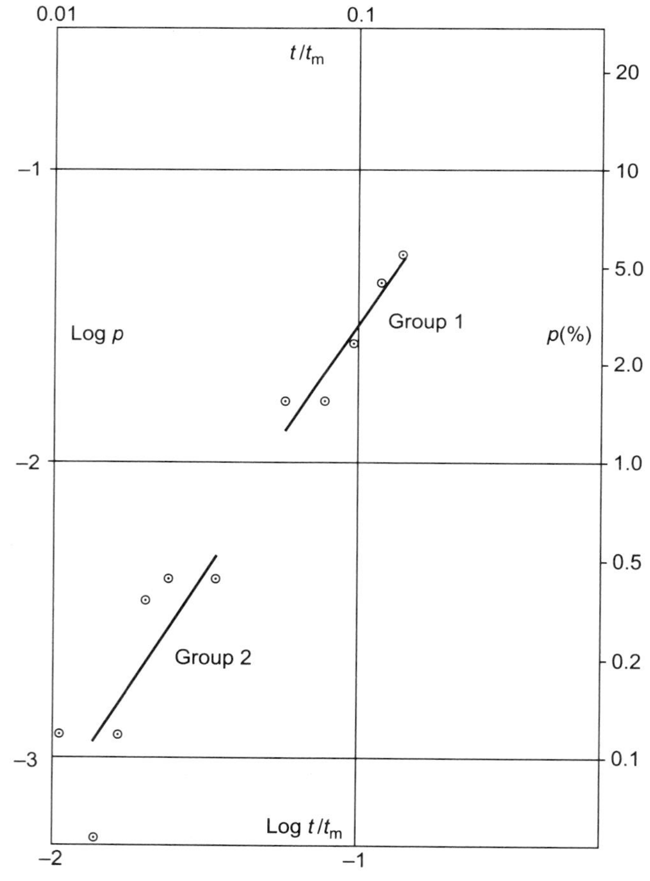

1.27 Weibull plot of tube failures in reformer furnaces: European survey.[14]

relatively high failure rate were all located on one site and were owned by the same company; the others were scattered over different locations in continental Europe with various owners. There was clearly a difference in the operating history of these two groups and they were therefore analysed separately, as indicated on the plot. In both cases the Weibull slope was similar, about 1.5. When this slope is equal to 1.0 the Weibull distribution is the same as the exponential distribution of chance failures. A figure of 1.5 indicates predominantly random failures with a modest increase of the failure rate over a period of time.

Four of the tube failures were due to casting defects but the remainder were likely to have been caused by accidental overheating. Reformer furnaces are fired by means of a large number of burners, and if these are not accurately adjusted for intensity and direction then local overheating is possible. It seems possible that operators in the group 2 plants, where the failure rate was an order of magnitude less than in group 1, were better at these and other control measures than those in group 1.

The investigation of reformer furnaces in the USA by Osman and Raziska of the Exxon Chemical Co,[15] also disclosed a majority of plants (23 out of 30) where the average failure rate fell below about 0.5% (as with group 2 of the European survey). Of those reporting substantial numbers of failures the Weibull slopes ranged between two extremes, those illustrated in Fig. 1.28 being typical of the extreme values. The plot labelled 'Plant 2' is very similar to group 1 of Fig. 1.27 and could likewise be interpreted as due to chance overheating of individual tubes. That labelled 'Plant 4', however, is the distribution which would be expected from tubes that had reached the end of their life and were about to fail *en masse* by creep-rupture. Table 1.11 gives details of the Weibull characteristics of furnaces or groups of furnaces having seven or more failures. These have high values of the slope, indicating that the tubes as a whole had reached the end of their useful life, three have relatively low slopes, typical of random types of failure, and three were at an intermediate level, suggesting a mixture of random and creep-rupture failures.

Furnace failures are rarely catastrophic; usually they constitute, at worst, a fire within a fire. Statistics about reformer furnaces have been included here, however, for two reasons. The first is that they demonstrate the predominant effect of the human factor in certain cases. The three plants in Group 1 of the European survey were located in north-east England, an area which in times past built much of the world's shipping. Thus there was no lack of indigenous skill. The furnace tube design was not significantly different from that of the others, yet the failure rate was ten times as high. The period in question was one in which

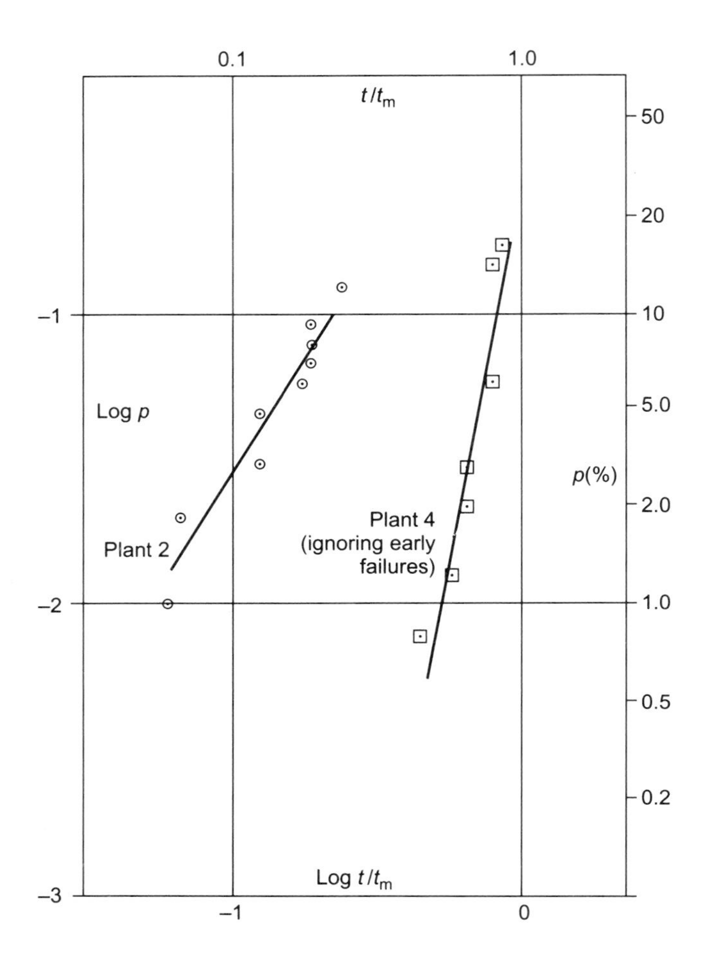

1.28 Weibull plot of tube failures in reformer furnaces: US survey of side- and bottom-fired units.[15]

management–worker relationships in England were at a low ebb, and it could be that the apparent lack of skill was associated with a negative attitude amongst the operations – 'won't do' rather than 'can't do'. However this may be, it is clear that differences of failure rates between these two groups of plants was not due to mechanical differences, but to differences in human behaviour.

The same applies to the US reformers surveyed by Exxon. Careful statistical work showed that design had no significant effect on the results. Nevertheless, some plants had failure rates that were an order of

Table 1.11 Results of Weibull plots of steam-methane reformer furnace tube failures

Plant	Furnace type	Weibull slope, *b*	Constant t_m (h)	Correlation coefficient[a]	Number of failures
Group 1	Kellogg	1.49	4.5×10^5	0.946	30
Group 2	Kellogg	1.58	2.6×10^6	0.873	7
2	Selas	1.64	2.84×10^5	0.975	9
4	Selas	8.42	1.07×10^5	0.983	7
7	Foster Wheeler	7.93	4.15×10^4	0.996	10
11	Selas	10.82	1.07×10^5	0.958	6
20	Selas	2.66	1.66×10^5	0.929	10
22	Foster Wheeler	3.19	1.30×10^5	0.987	11
23	Foster Wheeler	4.41	1.38×10^5	0.970	6

[a] When correlation coefficient = 1.0, all data points lie exactly on a straight line; when it is equal to zero, data points are scattered in a completely random fashion.

magnitude higher than others. However the majority of the US failures were by creep-rupture (68%) and in three plants there was evidence of premature failure; in other words the furnaces had been pushed too hard. Again, there was a difference in human behaviour, although in a positive rather than a negative direction.

The second reason for detailing these results is to demonstrate the value of the Weibull model. By means of a simple plot it is possible to predict the time for failure of any given proportion of components, and in some cases to diagnose the nature of the failures. In the reformer furnace tubes the plot distinguished between random and creep-rupture failures, without the need for any physical examination, simply because these two types of failure have a different probability distribution. Weibull analysis is not normally applicable to catastrophic failures, but may nevertheless be relevant by indicating the character of some types of mechanical breakdown.

Offshore operations

The discovery of oilfields offshore, in the Gulf of Mexico and in the Middle East, and later in other parts of the world including the North Sea, has led to the development of new techniques for exploration and drilling, and to the production and operation of specialised types of vessel and fixed platforms for drilling wells and for oil production.

Equipment

The various devices that are used for drilling and production offshore, and the way in which they have developed, are described in detail in Chapter 5. So far as losses are concerned, they are divided into two major categories: fixed and mobile. Fixed units are those that rest on and are secured to the sea-bed. They are used for the development drilling of a field, for production and for primary separation of oil, gas and water. Mobile units are those that from time to time move from one location to another, either under tow or under their own power. Jack-up units are barges fitted with legs that can be jacked down to the sea-bed and then further jacked to provide an elevated platform at a sufficient height to be clear of waves. Jack-ups are considered later as a separate category of mobile units.

Sources

A limited amount of information is published by various national bodies concerned with safety at work, for example the Health and Safety Commission in the UK, but the main source for offshore accident statistics is the World Offshore Accident Databank (WOAD) which is maintained by the staff of Bureau Veritas in Norway. This document is updated regularly and published every two years. While the coverage is worldwide, figures for countries with state-owned oil industries are not usually available and so are not included. Data about the number of workers on fixed offshore platforms are either unobtainable or unreliable. For mobile units these figures have been estimated from the known crew capacity of the vessel concerned. Therefore the annual fatality risk per million workers (quoted earlier in this chapter) applies to mobile units only. Bearing in mind these minor reservations, the WOAD material represents as good a set of accident data as could be expected for the period since 1970, when records began.

Bureau Veritas is a classification society; that is, it exists primarily to inspect and to classify ships, in the same way as Lloyd's Register of Shipping. Both these organisations have extended their inspection work to other fields, notably offshore operations, and Bureau Veritas is particularly well placed to maintain offshore records and to evaluate the nature of offshore losses and accidents.

Offshore losses

Figure 1.29 plots the numbers of individual fixed installations and the number of mobile units operating offshore. Detailed figures between 1970 and 1979 are not available in recent WOAD reports, so, when required, figures have been interpolated between these two dates.

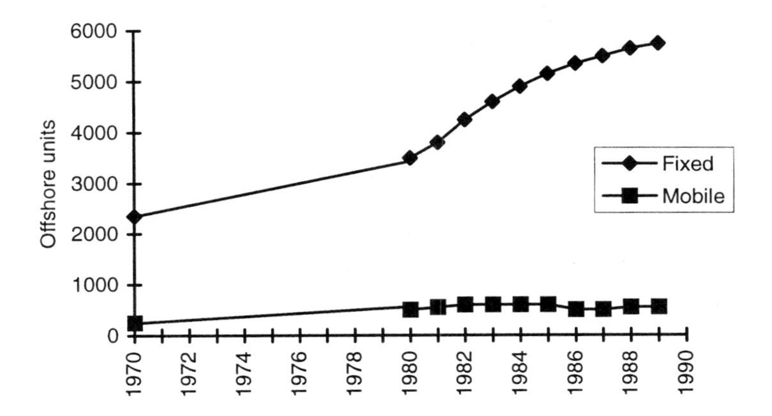

1.29 Numbers of offshore units 1970–90.[2]

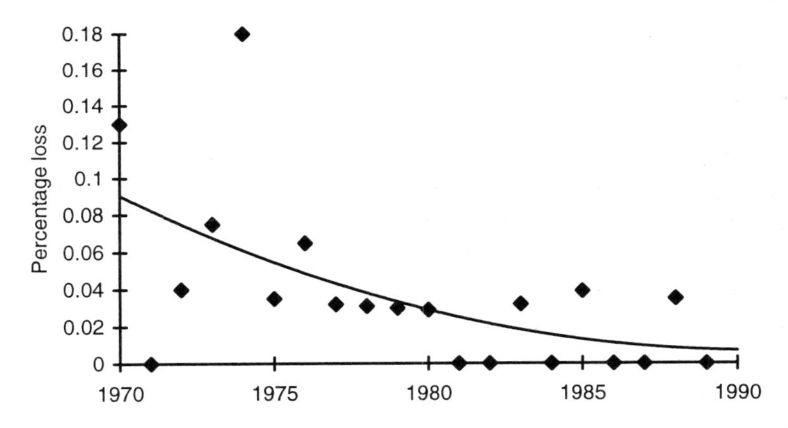

1.30 Percentage loss of fixed offshore units 1970–90.[2] Annual decrement = 7.1%

Figure 1.30 shows the percentage loss of fixed units for the two decades after 1970. There is undoubtedly a reduction in the loss rate over this period, but the amount of scatter is such that not much reliance can be placed on the figure given for the annual decrement.

A similar plot for mobile units (Fig. 1.31) shows a completely different picture. In the first place the mean loss rate is 26 times that for fixed units over the same period. Secondly, there is no significant change in this rate. The reason for relatively high and unvarying loss rate for mobile units in general will be considered later. In the meantime it is worth looking at the records for jack-up units as a separate category. On the face of it, these would appear to be subject to hazards over and above those associated with

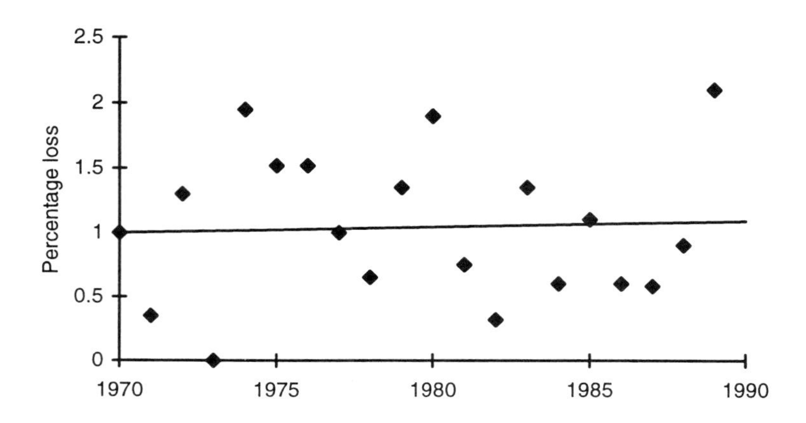

1.31 Percentage loss of offshore mobile units 1970–89. No significant change; mean = 1.05.

other mobile craft. A mechanical failure during a jack-up operation could be disastrous. This was the case with the *Sea Gem* where on a cold night in December 1965 such a failure did occur. The platform fell nearly 40 ft into the sea, was severely damaged, and sank. Thirteen men lost their lives. It is also possible for one or more of the legs to land on an unstable piece of sea-bed. In October 1985 the jack-up leg *Penrod 61* collapsed during a hurricane in the Mexican Gulf, and this was thought to have been due to soil failure beneath one of its three legs.

Table 1.12 lists the percentage of mobile units lost for the years 1980–89, together with the average loss during the two decades 1970–79 and 1980–89. They are divided into jack-up, nonjack-up and all mobiles. There is no discernible trend for the years 1980–89, but for jack-up units the average percentage loss for the 1980–89 period was significantly less than in 1970–79. Moreover the loss rate for jack-ups in 1980–89 was not significantly different from that for other types of mobile unit. It would seem that jack-up units are now only marginally less safe than other mobile units.

Figures for fatalities are presented as numbers per 100 unit years in Fig. 1.32 and 1.33. Those for fixed units (Fig. 1.32) show an apparent rise over the 1970–89 period. This is due to the inclusion of the casualties (165 in number) from the Piper Alpha disaster in 1988. These constitute over 40% of total deaths during the period and have a distorting effect on the result. If the Piper Alpha figures are excluded, the regression curve is virtually flat, as shown in the figure. Evidently loss of life is unrelated to loss of the units themselves. This is not unexpected because fatalities may

Table 1.12 Offshore mobile units: losses, 1970–89

Year	Jack-up			Other mobile units			All mobile
	Loss	Units operating	% loss	Loss	Units operating	% loss	Units % loss
1980	6	251	2.4	3	226	1.3	1.9
1981	2	364	0.7	2	240	0.8	0.7
1982	1	372	0.3	1	252	0.4	0.32
1983	5	353	1.4	4	231	1.7	1.5
1984	2	303	0.5	2	249	1.2	0.6
1985	5	385	1.3	2	251	0.4	1.1
1986	2	305	0.7	1	197	0.5	0.6
1987	3	324	0.9	0	190	0	0.6
1988	3	336	0.9	2	199	1.0	0.9
1989	8	337	2.4	3	199	1.5	2.1
1980–89	37	3352	1.1	20	2234	0.9	1.0
1970–79	22	1418	1.6	11	1674	0.7	1.3

be caused by accidents (falling loads for example) which would not normally affect the integrity of the structure.

Records for mobile units show no significant trend over the 20-year period, and again the average rate is over 20 times that for fixed units. Details are listed for the years 1980–89 in Table 1.13, together with the totals for 1970–79 and 1980–89. The number of personnel–years is also given in this table, and hence the calculated annual fatality risk per million workers. The apparent increase in risk from 9.8×10^{-4} to 12.11×10^{-4} is not considered to be significant.

Causes

WOAD lists the causes of losses in terms of events. These are given in sequence, the first being the initial event giving rise to the accident, followed by the resulting effects; thus, the initial event could be 'blowout', followed by 'explosion' and 'fire'. Twenty-one types of event are identified and defined.

Table 1.14 shows the number of total losses against the type of initial event. Also given are the corresponding percentage of all the losses. In this table all the categories customarily used by classification societies for shipping losses generally have been lumped together under the designation 'capsize, etc'. These categories include collision, contact, foundering, etc, as indicated below the table. In this way the incidents that might be considered to be a normal hazard in operating ships are separated from those peculiar to offshore oil exploration, drilling and production.

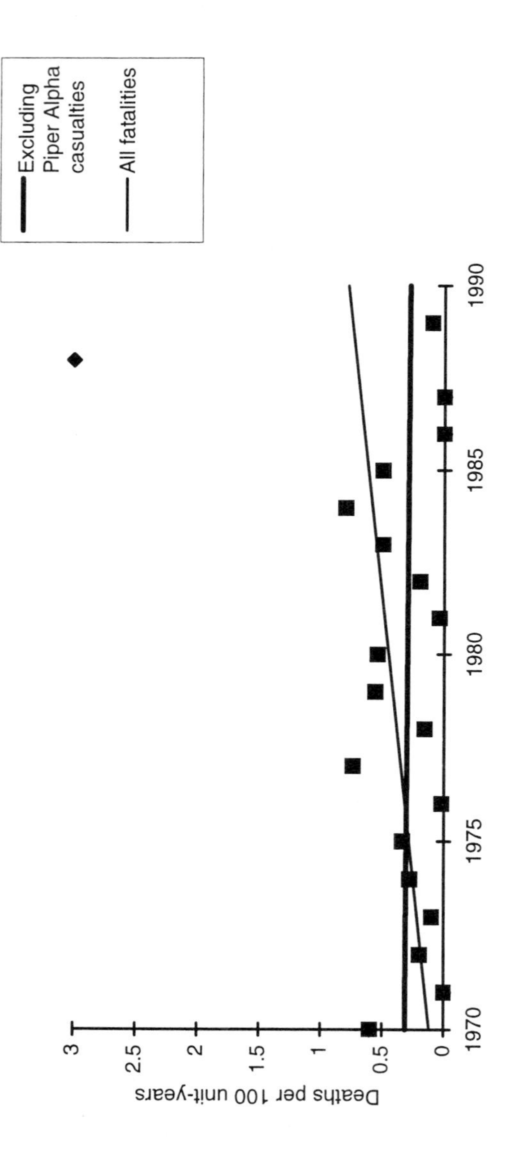

1.32 Fatalities per 100 unit-years, offshore fixed units, 1980–89. No significant change; mean = 0.45.

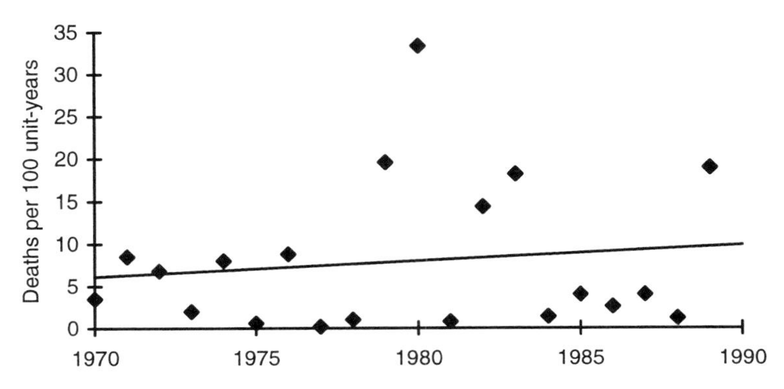

1.33 Fatalities per 100 unit-years, offshore mobile units 1980–89. No significant change; mean = 7.9.

Table 1.13 Loss of life from offshore mobile units, 1970–89

Year	Mobile units, worldwide		
	Deaths	Number of unit years	Deaths per 100 unit years
1980	155	477	32.5
1981	7	544	1.3
1982	92	624	14.7
1983	106	584	18.2
1984	6	634	0.9
1985	21	636	3.3
1986	7	502	1.4
1987	20	514	3.9
1988	7	535	1.3
1989	103	536	19.2
1980–89	524	5586	9.4
1970–79	203	3092	6.6

No. of personnel-years: 1970–79, 207 145; 1980–89, 432 800.
Risk per million personnel-years, mobile units: 1970–79, 980; 1980–89, 1211.

At first sight it might seem strange that 'Capsize, etc' should also apply to fixed units. However the majority of fixed units jackets are fabricated on shore, launched and then towed out to position, so they start life as a rather peculiar floating object.

The other (and most important, in the case of fixed units) cause of loss is blowouts. To quote WOAD a blowout is 'an uncontrolled flow of gas, oil or other fluids from the reservoir'. This is a hazard peculiar to drilling operations, and it is particularly difficult to control. This problem is discussed in Chapter 5.

Table 1.14 Initial event leading to total loss of offshore units, 1970–87

	Number	% of total
(a) Mobile units		
Capsize, etc[a]	31	42
Blowout	16	22
Structural failure	10	14
Towing accident	6	8
Explosion, fire	2	3
Other	9	11
(b) Fixed units		
Blowout	8	38
Structural failure	6	29
Capsize, etc[a]	4	19
Explosion, fire	2	10
Towing accident	1	4

[a] Includes capsize, collision, contact, foundering, grounding, leakage and list.

Structural failure makes a significant contribution to the loss rate. There is no information about the relative incidence of the various types of structural failure, but from evidence presented earlier in this chapter fatigue would be expected to make a major contribution. Certainly, the loss of the *Alexander L Kielland* resulted from a combination of low transverse ductility of the steel with fatigue cracking. Explosion and fire, rather surprisingly, come fairly low on the list.

The record for fatalities (Table 1.15) shows a very similar pattern. Structural failure comes high on the list for mobile units because of the *Alexander L Kielland* losses; excluding these brings it to the bottom of the table, as in the case of fixed units. Explosion and fire rate highly; and if the Piper Alpha figures are taken into account, head the list for casualty rates. Helicopter accidents also make a significant contribution.

The fatality rate per million persons at risk has already been quoted for mobile units. The numbers of workers on fixed units is not known other than for some individual areas. The crew of a large production unit may be 200 or more, but other platforms are unmanned or are manned only by a small number of workers during the day. An average figure of 100 is probably too high (the average for mobile craft is about 70) so we will guess at a figure between 25 and 50. If the losses on Piper Alpha are included, this would give an annual fatality risk of between 100 and 200 per million. Excluding Piper Alpha gives a figure of between 50 and 100. In other words the fatality risk in working on a fixed platform probably lies within the normal range for land-based industry, possibly at the higher

Table 1.15 Initial event leading to fatalities on offshore units, 1970–87

	All fatalities		Excluding *Alexander L Kielland* disaster		Excluding Piper Alpha losses	
	Number	% of total	Number	% of total	Number	% of total
(a) Mobile units						
Capsize, etc[a]	296	48	296	60		
Structural failure	126	20	3	1		
Blowout	66	11	66	13		
Helicopter accident	43	7	43	9		
Explosion, fire	25	4	25	5		
Crane accident	19	3	19	4		
Towing accident	14	2	14	3		
Other	29	5	29	5		
(b) Fixed units						
Blowout	72	20			72	36
Explosion, fire	223	61			58	29
Capsize, etc[a]	43	12			43	21
Helicopter accident	11	3			11	5
Structural failure	3	1			3	2
Crane accident	2	1			2	1
Other	12	2			12	6

[a] Includes capsize, collision, contact, foundering, grounding, leakage and list.

end if account is taken of the worst disaster. There is a difference, however; whereas for industry the fatality rates are falling, the underlying rate for fixed offshore units has not changed over the last two decades.

Boilers and pressure vessels

Equipment

Boilers come in all sizes, from the billy-can to the boiler of a power station, which may be as big as a cathedral. We are concerned here only with boilers that operate at elevated pressure, and are used for generating power or process steam.

There are two basic types of steam boiler, the fire-tube and the water-tube. Most small boilers are of the fire-tube or shell type, where the hot flue gas passes through tubes which are surrounded by water. The whole is encased in a shell. In effect it is a shell-and-tube heat exchanger with hot gas on the tube side and water and steam on the shell side; indeed in some petrochemical plants this type of heat exchanger is used to generate process steam. There are many configurations of fire-tube boiler, including of course the steam locomotive boiler, which has horizontal tubes. Others, such as the old donkey engine boiler, have vertical tubes.

A cross-section of a typical oil-fired water-tube power station boiler is shown in Fig. 1.34. Burners fire into a space which is surrounded by banks of vertical tubes joined sideways by fins to form a 'water-wall'. Water is fed into these tubes by bottom headers; as the water flows upwards it is converted into a steam–water mixture. The tubes discharge into a top header and then pass to a steam drum, where water and steam are separated, the water flowing back to the bottom header and the steam being passed on to superheaters and to the power turbines. Finally the steam is condensed and returned to the system. So far as safety is concerned, the main concern is with the steam drum, which has a number of inlet tubes attached to the upper part and one or more outlet nozzles along or near the 6 o'clock position connected to downcomers. These drums are large, heavy-walled vessels and in the past several have failed during hydrostatic testing (although none, so far as is known, has failed in service).

Similar boilers are used on large petrochemical plants to provide both electric power and process steam. In such cases there are piping connections between the steam drum and process units, and the number of nozzle connections to the steam drum is correspondingly increased.

Pressure vessels comprise all those vessels such as reactors and drums which operate under pressure but which contain fluids other than water and steam, and are not subject to heating by fire or burners.

Sources

There is no agency that keeps a count of losses of boilers and pressure vessels on an international scale. In individual countries the annual number of boiler explosions may be known; for example in the UK it is a requirement of the 1882 and 1890 Boiler Explosion Acts that such explosions be reported to the appropriate ministry. However, the total number of boilers in operation is rarely known with certainty. It is not as a rule obligatory to report pressure vessel explosions, and indeed there have been occasions when such information has been kept secret or at least not reported in the technical literature. The total number of pressure vessels is quite unknown, and cannot easily be estimated.

Therefore estimates of reliability for this type of equipment rely on surveys of sample populations, manufacturers' records, reports by inspection organisations and so forth. Eyers and Nisbitt[18] reported on the number of boiler explosions and fatalities therefrom in the UK between 1860 and 1960. Phillips and Warwick[19] surveyed the incidence of defects in pressure vessels and gave some information on reliability,

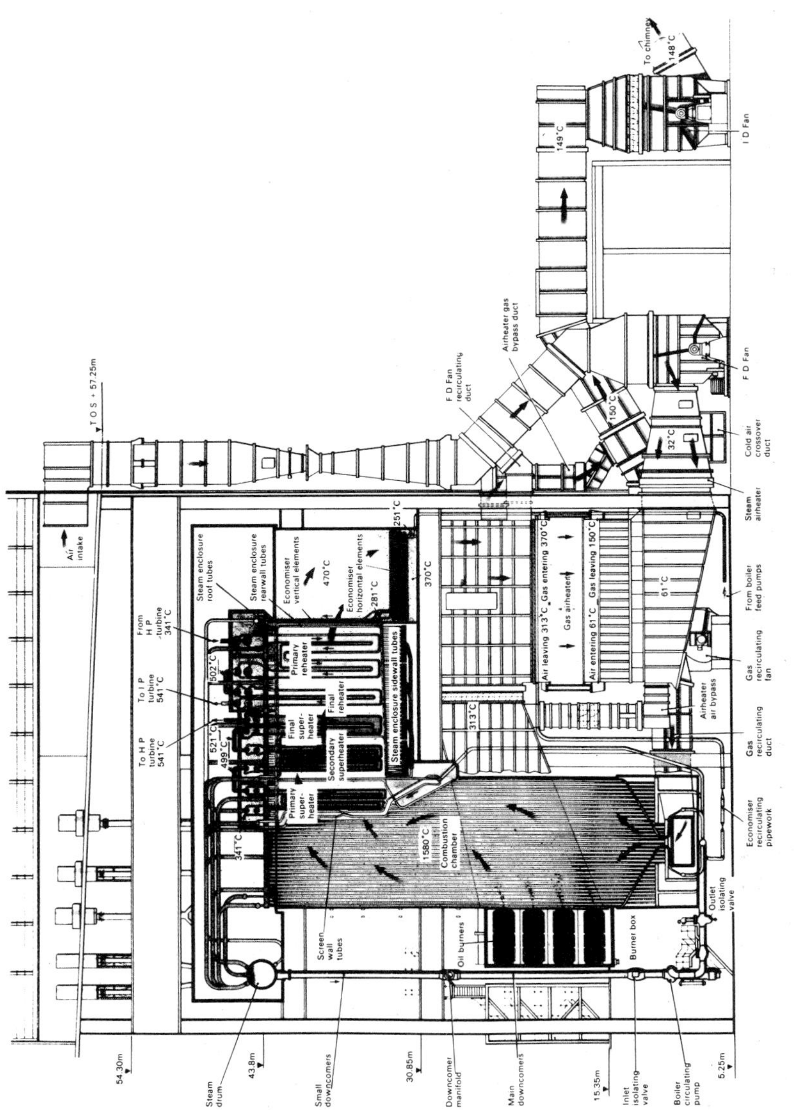

1.34 Oil-fired water-tube boiler for central power station.

Kellermann[20] of Technische Uberwachungs Verein (TUV) reported the failure rate of pressure vessels in Germany from 1958 to 1965. The present author[21] surveyed failures and their causes in the UK and Germany up to 1972, and Bush[22] provided a most wide-ranging survey, covering Germany, the UK and the USA up to 1975.

The work of Kellermann and Phillips and Warwick was motivated by the need to obtain an estimate of the reliability of nuclear pressure vessels. Subsequently two lines of activity have been followed to the same end: firstly, a development of techniques for predicting the conditions for leakage or catastrophic failure of a pressure vessel containing a defect of known size and shape, and secondly carrying out full-scale bursting tests on real defect-containing vessels. Means of predicting reliability are given in the ASME Code Sections III and XI and in various international and national codes and standards, such as the British document PD 6493. Burst tests have been carried out by the Netherlands Institute of Welding[23] and by the Technical Research Centre in Finland.[24]

Data

For historical trends, the main source is the document compiled by Eyers and Nisbitt, of the Associated Offices Technical Committee (AOTC). This organisation was originally the Manchester Steam Users Association, formed in the middle of the nineteenth century to carry out periodic inspections of boilers. Their records go back to 1860. Figure 1.35 shows the annual loss rate due to boiler explosions after 1890, based on successive five-year means. The 1890–94 period represented a peak for boiler explosions; prior to that time the rate was steady or increasing slightly. The record for injuries and fatalities due to boiler explosions shows a similar trend, as indicated in Fig. 1.36.

All these curves show a satisfactory downward trend. However, the number of boilers operating in the UK also fell during this period. At the start, individual machines in a factory were driven by a belt and pulley running on a common shaft, which in turn was driven by a steam engine. At the end, in 1960, virtually all such machines were powered by electric motors. The rate of decline in numbers of boilers for the period 1925–60 is indicated by Fig. 1.37. This shows the numbers of boilers certified annually by the AOTC. The total number of boilers operating would, of course, be larger; however, the AOTC would undoubtedly have been responsible for a substantial proportion of all certifications, and this proportion is unlikely to have changed very much between 1925 and 1960. Therefore, Fig. 1.37 may reasonably be taken to be representative of the actual trend.

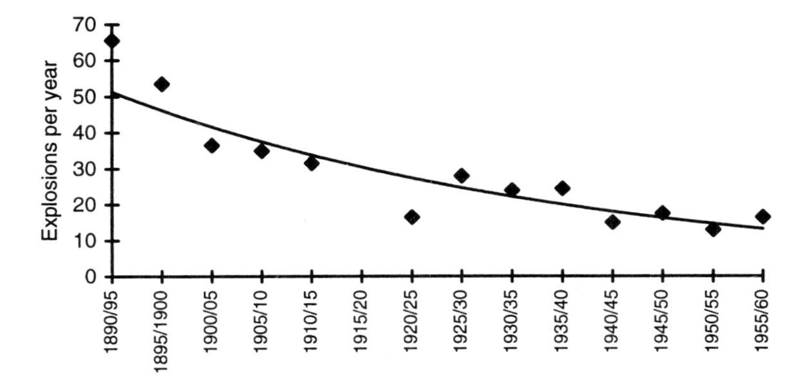

1.35 Annual rate of boiler explosions in the UK 1890–1960. Means of successive five-year periods.[18]

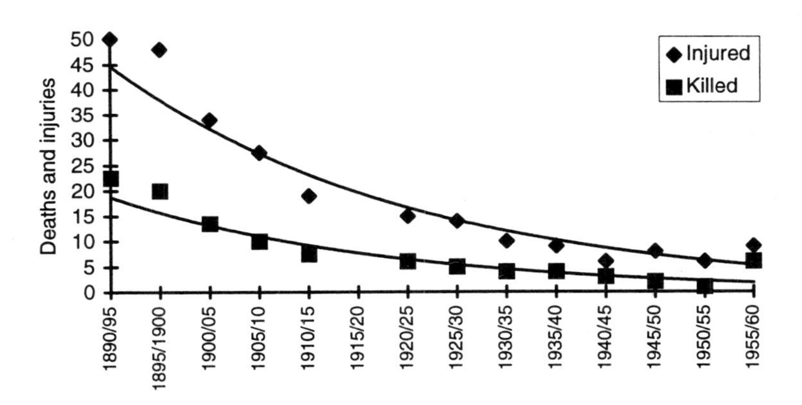

1.36 Annual rate of deaths and injuries due to boiler explosions in the UK 1890–1960. Means of successive five-year periods.[18]

Assuming this to be the case, the figures for boiler explosions, injuries and fatalities have been divided by the relevant boiler number shown in Fig. 1.37, to give the somewhat less satisfactory plots of Fig. 1.38, 1.39 and 1.40, for losses, injuries and fatalities respectively. The solid lines shown represent a least-squares fit to the data but there is no significant trend in any of the sets of figures. The mean values for percentage loss by explosion, and fatality or injury per 100 boiler–years are likely to be higher than the true figure, but this does not affect the conclusion that there was

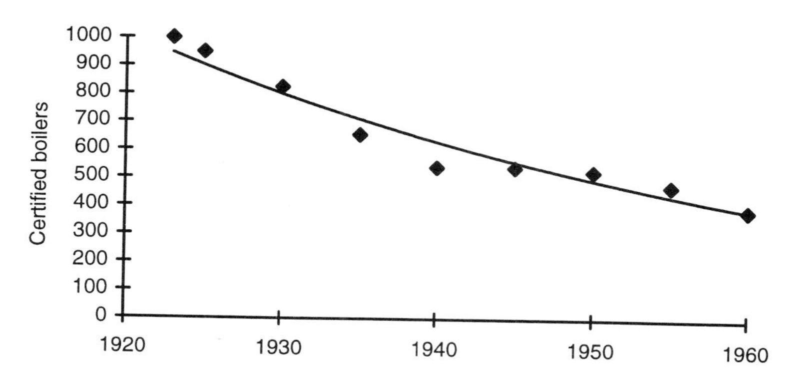

1.37 Number of boilers certified by the AOTC in the UK 1923–60.[18]

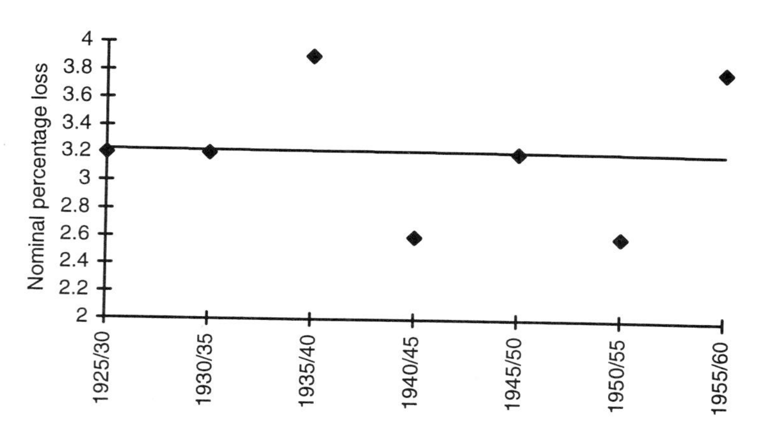

1.38 Nominal percentage loss of boilers due to explosions in the UK 1925–60. Mean = 3.2%.

no evidence to indicate an improvement in the safety of boiler operations in the UK during the period 1925–60.

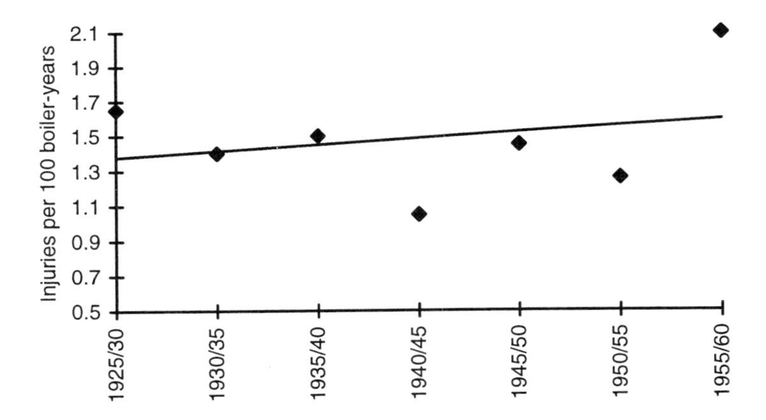

1.39 Nominal injury rate due to boiler explosions in the UK 1925–60. Mean = 1.5 per 100 boiler–years.

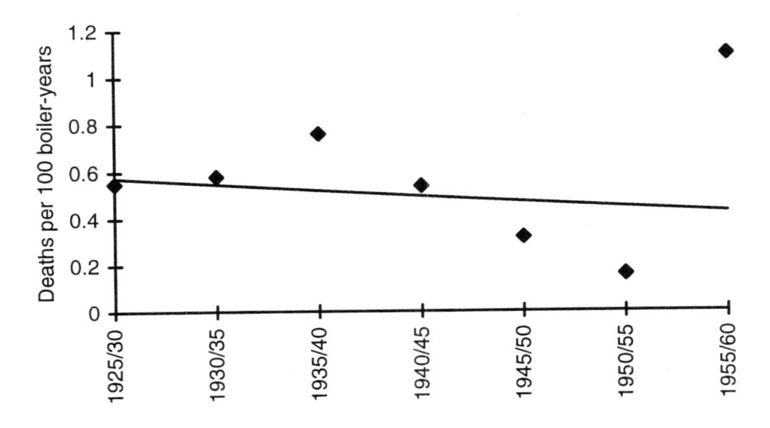

1.40 Nominal fatality rate due to boiler explosions in the UK 1925–60. Mean = 0.56 per 100 boiler–years.

Causes of failure

The causes of failure in service given by Phillips and Warwick and the accidents reported to US inspection agencies during 1973 are shown in Tables 1.16 and 1.17. These are in agreement about the main cause of trouble: maloperation, and in particular, allowing the water level to get too low. Eyers and Nisbitt in their earlier report came to the same conclusion, and do not mince words about those responsible:

Table 1.16 Causes of disruptive failures of boilers and pressure vessels in the UK during service, 1962–67[19]

Cause	Number	Percentage
Low water, etc	4	57
Manufacturing faults	1	14
Fatigue cracking	2	29

Table 1.17 Causes of accidents reported by US inspection agencies during 1973[22]

Cause	Number	%	Injuries	Deaths
(a) Power boilers				
Low water, etc[a]	689	72	5	0
Unsafe operation	97	10	1	0
Explosion/flareback	73	8	14	2
Shell rupture	11	1	4	0
Other	81	9	22	0
Total	951	100	46	2
(b) Heating boilers				
Low water[a]	968	91	0	0
Explosion/flareback	66	6	45	0
Unsafe practices	17	2	6	0
Shell rupture	6	1	0	0
Total	1057	100	51	0
(c) Unfired pressure vessels				
Unsafe practices	128	62	23	9
Air systems	23	11	2	0
Piping and hose	16	8	0	1
Jacketed kettles	8	4	0	1
LPG vessels	4	2	109	12
Other	38	13	3	1
Total	217	100	137	24

[a] Dry firing, low water and overheating.

an accident associated with low water is the direct result of neglect such as:

(a) The plant being under the control of untrained operators who do not know how to test the water gauges and who have little appreciation of the importance of their duties and who, in consequence, take wrong action in an emergency.
(b) Faulty operation and maintenance revealed by failure to make regular tests of the water gauges and by neglect in the maintenance of the various boiler fittings and associated equipment.
(c) Misplaced faith in the infallibility of automatic controls together with neglect in their regular testing and failure to observe a planned maintenance programme in connection with such apparatus.

Information about contemporary failure rates is lacking so that it is impossible to say whether or not this problem persists. However, it is interesting to note that in spite of increasing concern about safety and

Table 1.18 Causes of all failures (disruptive and non-disruptive) of boilers and pressure vessels, UK 1962–67[19]

Cause	Number	Percentage
Cracks	118	89
Maloperation	8	6
Manufacturing defect	3	2
Corrosion	2	2
Creep	1	1

Table 1.19 Causes of cracking which resulted in failures of boilers and pressure vessels, UK 1962–67[19]

Cause	Number	Percentage
Fatigue	47	40
Corrosion (including stress-corrosion and corrosion-fatigue)	24	20
Manufacturing defects	10	8
Unknown	37	32

quality there appeared to be no improvement from 1925 to 1960, a period of 35 years. And the problem was still there when Bush made his survey in 1975.

Causes of both disruptive and non-disruptive failures of boilers and pressure vessels are listed in the Phillips and Warwick review. These are shown in Table 1.18. Ninety-five per cent of the failures were non-disruptive, and 89% of these were due to cracks, mostly found during routine inspection. Nearly all the cracks whose cause could be identified were due either to fatigue or to corrosion-assisted fatigue (Table 1.19).

Boiler explosions that cause injuries and fatalities are characteristic of shell boilers. Water-tube boilers, particularly those used in power generation, suffer occasionally from tube rupture, but even when these result in quite violent explosions, they are generally contained within the massive structure and are unlikely to do harm to human beings. They can, however, cause serious disruptions and financial loss. For example the bursting of a water-wall tube may require the boiler to be shut down and cooled, scaffolding to be erected, the tube to be repaired, non-destructive and hydrostatic testing to be carried out, scaffolding to be removed and plant to be restarted; a very time-consuming operation. Bearing in mind that a large utility boiler may have 50 miles of tubing in the furnace zone and a similar length in the superheater, reheater and economiser coils, it is necessary to achieve a high level of reliability.

The types of failure to which such tubes may be subjected are described and illustrated in two publications, one by French[25] and a later book by

Table 1.20 Types of failures in water-tube boilers[25]

	Percentage
(a) Mechanical (81% of total)	
Overheating	75
Graphitisation	5
Fatigue	4
Erosion	4
Weld failures	3
Swages	3
Tube ties, legs	3
Other	3
(b) Corrosion (19% of total)	
Impure boiler feedwater	37
Hydrogen damage	20
Fuel ash corrosion	19
Oxygen pitting	11
Stress corrosion cracking	8
Caustic attack	5

Port and Herro:[26] French also made an analysis of failures submitted by customers of the Riley-Stoller Company (manufacturers of water-tube boilers) over a period of more than 20 years. These were considered, reasonably enough, to be a representative sample of such losses in water-tube boilers generally. The causes are listed in Table 1.20.

Most tube failures, as would be expected, result from overheating. There are many factors that give rise to the local overheating of tubes: steam blanketing, deposits, low flow rates, too-rapid start-up, load cycling and so forth. Low water is a rare cause. As a rule, a single tube bursts in one place, but sometimes banks of tubes may be affected, particularly if the overheating persists over an extended period. Graphitisation is shown as the second commonest cause of bursting. When carbon steel is held for long periods at temperatures above 450 °C iron carbide, which is present in the as-rolled material, breaks down to form nodules of graphite. Normally these nodules are scattered and do not have a seriously damaging effect, but in the vicinity of welds they may form chains and cause laminar weakness.

Most corrosion damage results from upsets in the water treatment. Boiler feed water is treated to remove solids, to reduce dissolved salts to a low level and to remove dissolved oxygen. If any of these operations fails, or if there is accidental contamination by carry-over or leakage of cooling water into the system, then deposits may form or tubes may be corroded. They may also be attacked externally by corrosive ash from the fuel.

Finally, the location of failures is analysed in Table 1.21. Tube wall temperatures are highest in superheaters and reheater coils, so it is not surprising to find that most failures are located in these two areas. At the

Table 1.21 Location of failures in water-tube boilers[25]

	Percentage
Superheaters	45
Water walls	29
Reheaters	14
Economisers	5
Roof	2
Floor	1
Others	4

Shipping

Source

Casualties of world shipping are recorded by Lloyd's Register of Shipping, which has published an annual statistical summary since 1891.[27] This organisation was set up to assist underwriters to insure ships and their cargo; it surveyed ships and classified them according to the condition of the hull and equipment. The results are set down in Lloyd's Register, the first copy of which was published in 1760. The register was eventually expanded to cover shipping worldwide, and the casualty returns are close to a complete record for the world fleet other than naval and military shipping, and countries that do not make such information available. It covers all classes of commercial shipping that have a displacement of more than 100 tons, so it may be taken as a large and representative sample.

The losses are classified as either foundered, missing, fire/explosion, collision, wrecked/stranded or lost. In relatively recent years the category 'contact' was added to the list; this refers to a damaging encounter with a natural or man-made object, other than the sea bed or another ship. Fatalities have been reported since 1978.

Data

The tonnage of world shipping listed in Lloyd's Register is plotted at ten-yearly intervals between 1890 and 1980 in Fig. 1.41. More detail is shown for the period 1980–92. The tonnage of world shipping reflects the volume of world trade, and would be expected to follow a similar course to the amount of oil produced; nevertheless, the similarity between this plot and that for oil shown in Fig. 1.20 is quite remarkable.

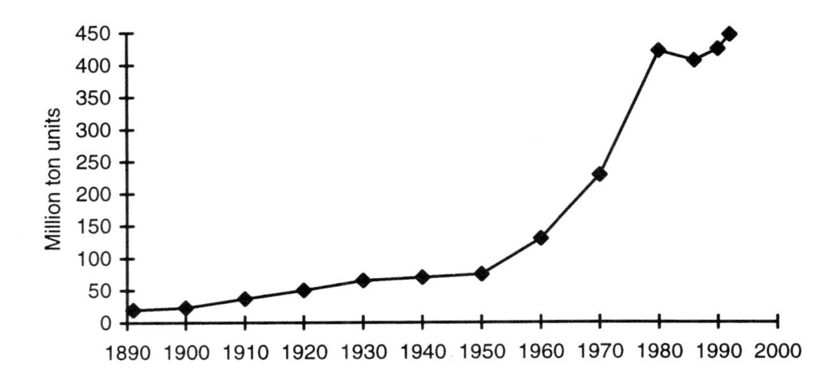

1.41 World shipping tonnage in million ton units 1891–1992.[27]

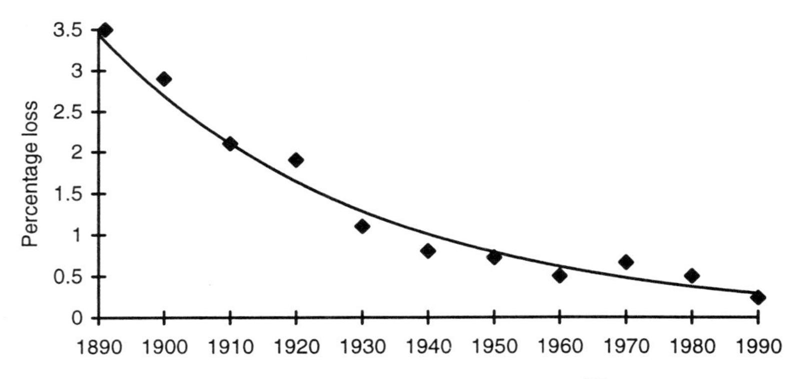

1.42 Annual percentage of ships lost worldwide 1891–1990.[27] Annual decrement = 2.4%.

Losses are recorded in terms both of numbers of ships and as tonnage. The corresponding percentage loss is shown for ship numbers in Fig. 1.42 and 1.43. The first curve encompasses the period 1891–1990, once again at ten-yearly intervals. The correlation is good and the annual decrement is 2.4%. Fig. 1.43 covers the period 1970–92 on an annual basis.

Fig. 1.44 is the equivalent plot for the percentage loss of tonnage, and is very similar to that for numbers. The improved reliability of ships indicated by all these figures is due to a number of factors, which will be discussed in a later chapter. However, as noted earlier, one important influence was the change from sail to steam. In 1891 about one-fifth of the shipping fleet consisted of sailing ships, which had an annual loss rate of about 5%, compared with 3.5% for steam ships. Sailing ships disappeared completely from the shipping fleet just a few years later.

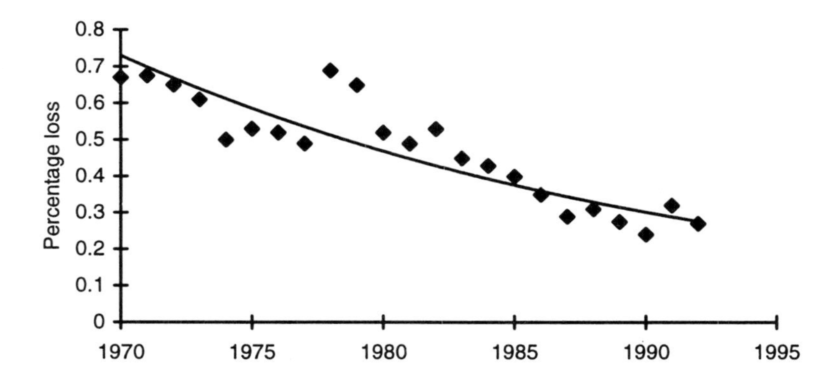

1.43 Percentage of ships lost 1970–92.[27] Annual decrement = 4.3%.

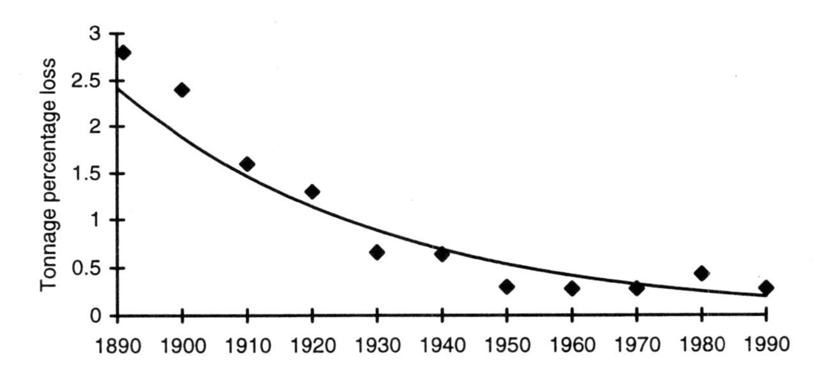

1.44 Annual percentage of tonnage lost. Shipping worldwide 1891–1990.[27] Annual decrement = 2.5%.

Fatality rates for 1978–92 are plotted in Fig. 1.45. There is a lot of scatter in these figures but the indication of a fall in fatality rates is consistent with long-term trends.

The proportion of fatalities and ship losses that fall into Lloyd's various categories are given in Fig. 1.46. Most fatalities, as would be expected, occur in ships that founder, and the highest proportion of ship losses fall in the same category. Tonnage losses are more or less equally divided between foundered, fire/explosion, wrecked and lost. However, if the figures for 'foundered' and 'lost' are merged, the division is more or less the same as for ship numbers.

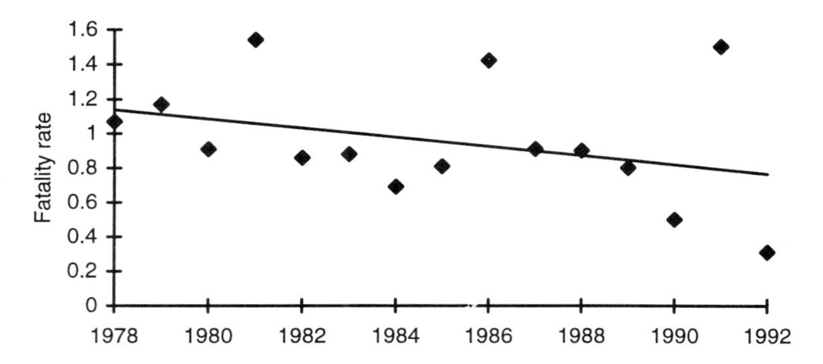

1.45 Shipping fatality rate per 100 ship–years 1978–92. Annual decrement = 3.96%.

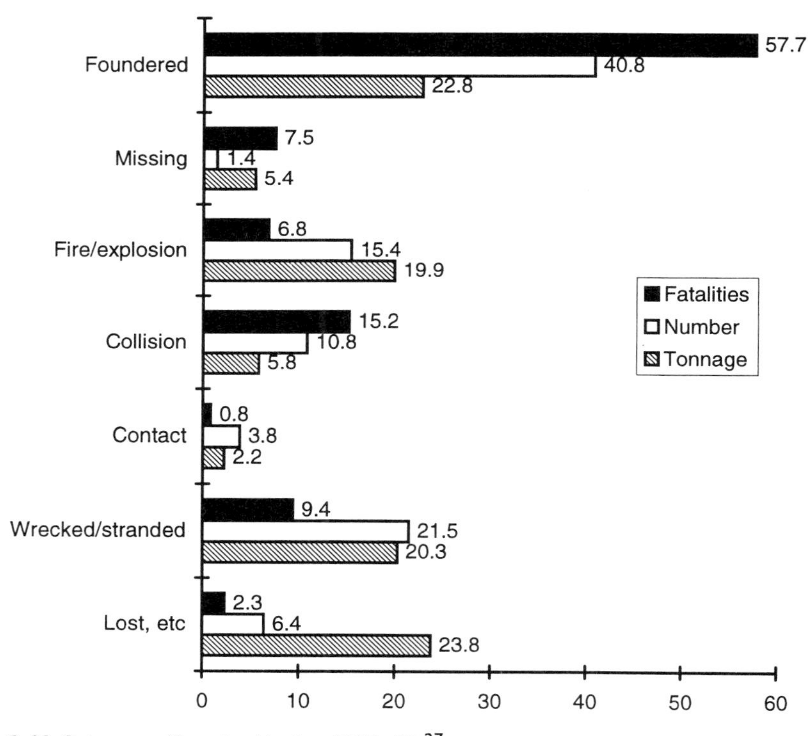

1.46 Category of loss in shipping 1983–92.[27]

Comparisons

Offshore

Annual fatality rates per million at risk in the UK are listed in Table 1.22. The worldwide figure for offshore mobile unit fatality rates in the 1980–89 period is 1211, similar to the UK figure, and the annual decrement (percentage reduction in the rate) is zero. If we take the 1989 figure for death by road accident (98 per million of the total population) as being a generally accepted level of risk, then all the figures listed in the table would fall in the acceptable range with the exception of offshore operations, which has a rate one order of magnitude higher. In an attempt to elucidate this problem WOAD records have been used to separate the fatalities in mobile units due to the type of accident to which shipping in general is subject (capsize, collision, founder, etc) from the remainder. Explosion and fire, although listed as a category in Lloyd's casualty reports, is included with the non-shipping type of losses because of the greatly enhanced risk from this cause in offshore units generally. The results for the years 1970–87 are presented in Table 1.23.

The first point to be noted about these figures is that of the total number of deaths due to shipping-type accidents, which amounted to 293, no less than 246 (84%) resulted from the loss of three units, all of which foundered in severe weather. If these losses are removed from the total, then the fatality rate per 100 unit-years falls to 0.6; exactly the same as that obtained for shipping generally when the Philippines ferry-boat disaster

Table 1.22 Annual death-rate per million persons at risk throughout the year 1985, and annual decrement for the period listed

Industry (Employees except where indicated)	Annual death rate, per million at risk	Annual decrement %	Period
All industries:			
All casualties	20	4.3	1961–91
Excluding Piper Alpha casualties	19	4.5	1961–91
Self-employed	26[a]	none	1982–92
Manufacturing	21	3.4	1982–92
Agriculture and coastal fishing	77[a]	none	1982–92
Energy and water supply	85[a]	none	1982–92
Energy and water, excluding Piper Alpha	77	3.0	1982–92
Construction	99	1.3	1982–92
Offshore oil and gas	1250		1987–90

[a]Mean for the 1982–92 period.

Table 1.23 Mobile offshore units, annual death-rate per 100 unit-years worldwide, 1970–87: shipping-type fatalities compared with those due to other causes

Year	Shipping-type fatalities		Fatalities due to other causes	
	Number	per 100 unit-yrs	Number	per 100-unit years
1970	0	0	5	3.6
1971	0	0	24	13.7
1972	0	0	14	6.7
1973	0	0	4	1.7
1974	19	6.9	3	1.1
1975	1	0.3	3	1.0
1976	7	2.0	23	6.7
1977	1	0.3	0	0
1978	2	1.5	1	0.25
1979	81[a]	18.3	6	1.4
1980	1	0.2	154[b]	28.3
1981	1	0.2	1	0.25
1982	84[c]	13.5	5	0.8
1983	81[d]	13.9	25	4.3
1984	0	0	7	1.1
1985	15	2.36	12	1.8
1986	0	0	13	2.6
1987	0	0	20	3.9
Total casualties	293		320	
Mean, all fatalities		3.8		4.3
Mean, excluding major losses		0.6		2.7

[a]*Ocean Ranger* 81.
[b]*Alexander L Kielland* 123.
[c]*Glomar Java Sea* 84.
[d]*Bohai II* 72.

is ignored. In two of these large mobile unit losses, the subsequent inquiry found that the fatalities could have been avoided or reduced by timely evacuation; in other words there might have been an unwise acceptance of risk by the crew. However this may be, it is clear that offshore craft are vulnerable to severe weather conditions, and this fact makes a substantial contribution to the overall fatality risk.

The non-shipping type of event (Table 1.24) includes anchor failure, blowout, crane accidents, explosion, fire, helicopter accidents, spillage, structural damage and towing accidents. These are all problems that are typical of offshore operations, and their hazard is increased by the confined space. Whereas in land-based hydrocarbon processing plant the operators are either widely dispersed or protected in a control room, on an offshore unit a major incident is likely to affect the complete crew. Moreover there are additional types of risk; blowout, helicopter accidents and towing accidents for example. So it is not surprising that the fatality

Table 1.24 Causes of non-shipping fatalities in mobile offshore units, 1970–87, excluding those due to the *Alexander L Kielland* disaster

Type of accident	Number killed	Percentage of total
Blowout	66	33
Helicopter accident	43	21
Explosion/fire	25	13
Crane accident	19	10
Towing accident	14	7
Structural failure	3	2
Machinery failure	2	1
Other	27	13

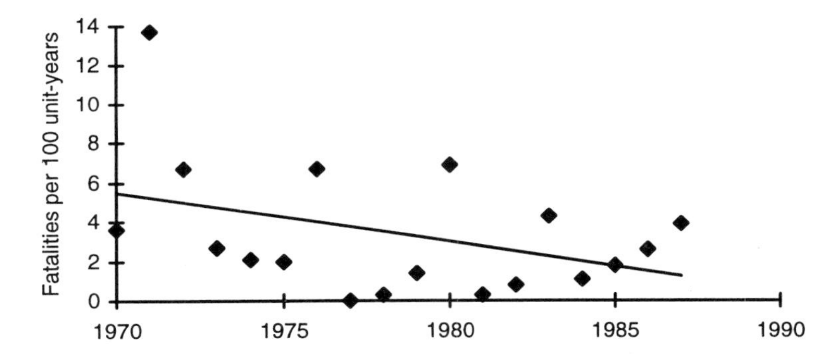

1.47 Fatalities in mobile offshore units due to accident other than those characteristic of shipping worldwide 1970–87 (excludes *Alexander L Kielland* casualties). Annual decrement = 4.7%.

and loss rates should be relatively high, although there is undoubtedly scope for improvement. In fact it seems likely that the non-shipping type fatality rate is falling. Figure 1.47 is a plot showing this rate for the years 1970–89. The correlation is low due to the periodic upward scatter of data points, but there does appear to be a modest downward trend. When these figures are put together with those for shipping-type accidents, this downward trend is completely masked by the three large losses in 1979, 1982 and 1983 (Fig. 1.33).

To summarise: it is possible to discern two more or less equally important factors contributing to the high fatality rate in offshore mobile units in the 1970–87 period; the vulnerability of offshore craft to severe weather, and the inherent risk in dealing with potentially inflammable and explosive substances in a confined area. In the first category there was no decrease in the fatality rate, but there were indications that improvements in evacuation procedures could change this situation. In the second category there is scope for improvement but it has to be recognised that there will always be special problems such as blowouts that make offshore work an intrinsically more dangerous occupation than, say, the manufacture of boots and shoes.

Hydrocarbon processing

It is important to try to pinpoint the reasons for the increasing rate of financial loss due to accidents in hydrocarbon processing plant. Mahoney[8] provides some leads. The increase in the total number of losses is attributed to better data-gathering and to increasing average age of plant. The greater magnitude of losses is put down to the larger size of individual units, to the change from multiple train to large single train plants, and to the increased concentration of equipment per unit area of the plot plan.

In an attempt to shed more light on this problem, the losses recorded for the period 1963–92 have been categorised in terms of the first event leading to the accident. Four categories were chosen: vapour cloud explosion, internal explosion, fire and natural hazards. Rather surprisingly, all of the losses fell into one or other of these slots. 'Internal explosion' covers all those cases where some part of the equipment was subject to excessive pressure and, as a result, exploded. Some excess pressures were due to process upsets (water coming into contact with hot oil, for example) or operator error. In some instances internal explosions were followed by vapour cloud explosions, but these have been included in the first category. 'Fire' includes all those conditions where a leak was followed by immediate ignition. Vapour phase explosions occurred where a quantity of gaseous or vaporised hydrocarbon formed a cloud which expanded until it reached a source of ignition. Natural hazard comprised two cases of lightning strikes, one earthquake and one hurricane. The results, for successive five-year periods, are given in Table 1.25, from which it will be obvious that the proportion of losses due to vapour phase explosions increased dramatically during the 30-year period, in terms both of numbers and capital cost.

The dollar loss due to vapour phase explosions, internal explosion and fire as a proportion of the nominal value of oil and gas produced is shown for the six successive periods in Fig. 1.48. This shows that losses due to vapour cloud have increased rapidly while those due to fire and internal explosion show only a modest rise.

Thus it would seem that the increased capital loss relative to the value of oil and gas produced was largely due to vapour cloud explosions. From Table 1.25 it may be found that the increase was due to both larger numbers and a higher average loss; in other words, vapour cloud

Table 1.25 Events leading to major loss in hydrocarbon processing plant[a]

Period	Vapour cloud explosion	Internal explosion	Fire	Natural hazard	Total hazard
(a) Number of losses					
1963–67	1 (17)	2 (33)	2 (33)	1 (17)	6
1968–72	1 (13)	4 (50)	3 (37)	0 (0)	8
1973–77	6 (38)	4 (25)	5 (31)	1 (6)	16
1978–82	7 (35)	5 (25)	7 (35)	1 (5)	20
1983–87	6 (33)	6 (33)	6 (33)	0 (0)	18
1988–92	17 (53)	9 (28)	5 (16)	1 (3)	32
Total	38	30	28	4	100
(b) Capital loss (US $ million)					
1963–67	76.6 (31)	62.5 (25)	98.8 (40)	10.9 (4)	248.8
1968–72	17.7 (6)	157.1 (49)	145.4 (45)	0 (0)	320.2
1973–77	354.8 (42)	105.5 (12)	369.2 (44)	15.6 (2)	845.1
1978–82	436.2 (50)	144.3 (16)	281.4 (32)	18.2 (2)	880.1
1983–87	462.8 (38)	375.1 (31)	369.3 (31)	0 (0)	1207.2
1988–92	2045.5 (75)	403.0 (15)	206.5 (8)	65.7 (2)	2720.7
Total	3393.6	1247.5	1470.6	110.4	6222.1

[a] Percentages in brackets.

explosions increased both in frequency and in the amount of damage caused.

In the hope of finding the source of the vapour clouds, the records were further analysed to determine the initial failure type. Table 1.26 gives the results. Piping failure was by far the largest single cause. In a high proportion of cases (Table 1.26) the failure mechanism was unknown or unrecorded; for the remainder, the causes were many and various. The bursting of elbows due to thinning by corrosion/erosion was marginally the most frequent. This type of failure gave rise to one of the worst fatal accidents in hydrocarbon processing plant (not recorded in the Marsh and McLennan survey; the capital loss was relatively small). This occurred in a CO_2 removal system where the gas was absorbed by hot potassium carbonate solution. Such units are fabricated from carbon steel, and an inhibitor is added to the carbonate solution to prevent corrosion. However, under conditions of high flow rates or turbulence, corrosion may still occur. In the unit in question a bend was positioned downstream of a stainless steel valve. To save money, this control valve was made two pipe sizes smaller than the piping, so it generated a jet of carbonate solution which impinged on the outer radius of the bend. By great misfortune this bend failed in such a way that the hot carbonate solution inundated the control room, killing all those within.

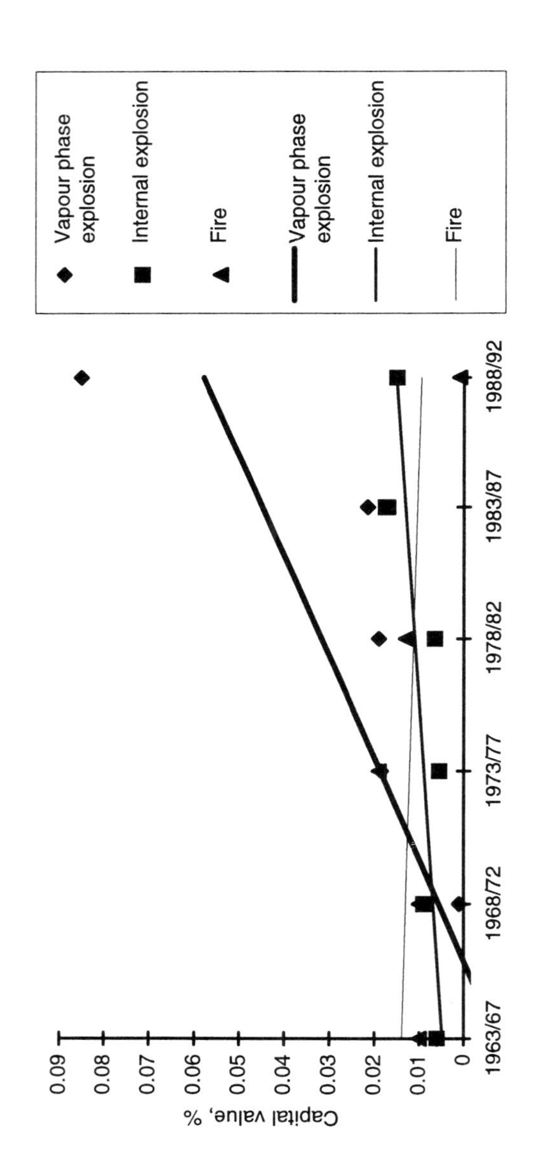

1.48 Capital value of large process plant losses as a percentage of the value of hydrocarbon produced due to: (a) vapour phase explosions; (b) internal explosions; (c) fire.

Table 1.26 Vapour cloud explosions and piping failures, 1963–92

	Number	Percentage
(a) Event leading to vapour cloud explosions in hydrocarbon-processing plant		
Piping failure	23	60
Operational upset or error	6	16
Contractor error	1	3
Spillage	1	3
Vessel failure	1	3
Unknown	6	16
(b) Cause of piping failure		
Erosion/corrosion of elbow	3	13
Expansion joint failure	2	9
Leak	2	9
Weld failure	2	9
Brittle failure	1	4
Fatigue	1	4
Valve failure	1	4
Corrosion of pipe	1	4
Unknown	10	44

Vapour cloud explosions usually result from the sudden release of inflammable gas, the worst case being a full-bore rupture of a pipe carrying (for example) ethylene at high pressure. During the post-1945 years, such accidents have become an increased risk owing to the increasing proportion of process plant that is handling gaseous hydrocarbons. The amount of new natural gas that is processed has increased dramatically, as has the production of ethylene and ammonia, for which a methane or naphtha feedstock is required. By itself this factor does not provide a complete explanation of the increased losses, however. The greatest increase in dollar loss occurred in the 1987–91 period, when construction of new petrochemical plant was almost at a standstill, refinery plant capacity fell and natural gas production increased at a somewhat lower rate than previously.

A second predisposing factor is the use of higher pressure both in pipelines and in process equipment. In the case of pipelines this has resulted in the use of higher tensile steel for the pipe. Such materials are more prone to weld defects and have an inherently lower ductility than plain carbon steel. However, there is not much evidence that the use of high tensile steel has been the cause of an increased incidence of failure. The higher operating pressures are more likely to have the effect of increasing the mass of fluid released.

To summarise, the increased losses in hydrocarbon processing plant are due primarily to an increase in the incidence and severity of vapour cloud explosions. To some degree this is due to an increase in the proportion of gaseous hydrocarbon being handled. However, during the past decade there has been little new plant construction, yet the losses have continued to increase. It has been suggested that the age of plant may be a contributory factor, but factual evidence to this effect is lacking.

The predominant cause of the formation of vapour clouds is the rupture of piping. It follows that the most effective means of reducing the incidence of vapour cloud explosions would be to improve the integrity of piping systems. Improving the quality of design, manufacture, inspection, repair and replacement of pipework could achieve this end, but it would require a considerable effort and considerable expenditure.

It is most fortunate that the large explosions and fires that afflict the onshore hydrocarbon industry are not accompanied by a corresponding loss of life. Because of this fact, however, there is little public awareness of the problem, and the incentive to take effective remedial action is correspondingly reduced.

General comparisons

The fatality rate per 100 unit-years for aircraft, offshore units, steam boilers and shipping is compared in Table 1.27. The year 1985 was chosen as a suitable date for comparison because more data are available for this period. Where there is a decreasing fatality rate the figure is taken from the least-squares line fitting the plot of ln (death-rate) versus time. Where there is no significant correlation then the figure quoted is the mean rate for the period.

Table 1.27 Annual death-rate per 100 unit-years worldwide during the year 1985 and decrement for the period listed

Item	Death-rate per 100 unit-years	Annual decrement	Period
Jet aircraft	6.5	5.9	1970–89
Offshore			
fixed[a]	0.5	None	1970–89
mobile[a]	7.9	None	1970–89
Steam boilers[b]	0.6	None	1925–60
Shipping			
all	1.0	None	1978–92
excluding Philippines ferry-boat disaster	0.6	4.0	1978–92

[a]Average for period.
[b]Estimate for UK only: average for period.

Once again, offshore mobile units head the list. The figure for jet aircraft is not of much significance because the main concern is for passenger safety and passengers only occupy their seats for short periods of time. A better figure relates fatalities to a number of departures, as noted earlier in this chapter.

Annual percentage losses of units worldwide are listed in Table 1.28. For aircraft, steam boilers and shipping the figures are very similar; offshore fixed units had a loss rate above an order of magnitude lower and mobile units about an order of magnitude higher.

Finally, an attempt has been made (Table 1.29) to determine the percentage of accidents that may be assigned to the main categories discussed earlier. Many of the causes listed in the source documents cannot, of course, be unequivocally fitted into a single main category. For example, a ship or offshore mobile unit that founders may do so primarily because it has been overwhelmed by high seas (natural hazard) or because of cracks or incorrect design (mechanical failure). In such cases the

Table 1.28 Annual percentage loss of units worldwide, 1985

Item	Loss rate (%)	Annual decrement (%)	Period
Jet aircraft	0.25	5.2	1965–92
Offshore units:			
fixed	0.03	6.1	1970–90
mobile[a]	1.1	None	1970
Steam boilers[b]	0.2	None	1925–1960
Shipping	0.3	2.4	1891–1990

[a]Average for the period.
[b]Estimate for UK only, average for the period.

Table 1.29 Cause of total loss: percentage of total

Causes	Aircraft	Hydrocarbon processing	Offshore units: Fixed	Offshore units: Mobile	Offshore units: All	Shipping: Number	Shipping: Tonnage
Mechanical failure	19	41	34	16	19	27	22
Natural hazard	5	5	49	47	47	41	37
Operational error	64	29	17	27	24	32	41
Sabotage	3	4	—	—	—	—	—
Unknown, other	9	21	—	10	10	—	—

number of losses has been divided equally between the two most likely categories; for example, in the case of shipping, 'foundered' has been split equally between mechanical failure and natural hazard. Details are given in Table 1.30.

Sabotage has been included because both aircraft and hydrocarbon processing plant are favourite targets for terrorists. Ships are rarely attacked in this way; they may, however, be sabotaged by scuttling. One case, for example, was that of a cargo ship in the Mediterranean sea which was scuttled to avoid being captured by the Turkish Navy. The cargo which it was carrying consisted of three tons of heroin. Other ships have been scuttled in order to collect the insurance on a non-existent cargo. Unfortunately no reliable figures for this type of loss are available.

Natural hazards are the most frequent cause of loss of offshore units, less important for ships, and have a low rating for aircraft and hydrocarbon processing units. This makes sense, because blowouts, which are an important cause of loss, only affect offshore operations. Aircraft, and particularly jet aircraft that are covered in the Boeing survey, are largely independent of weather conditions except when taking off and landing.

General summary

Most of the activities surveyed here show a consistent decline in the fatality and loss rate. In the case of shipping this decline has persisted since records began in 1891, whilst in other fields it also goes back to the first statistics. The decline is generally exponential such that the loss rate decreases each year by a percentage of the previous year's figure. On a smoothed-out curve this decrement lies typically in the range 2–5%.

For certain industries this is not the case, and the loss and/or fatality rates have not changed significantly, at least in recent years. Included in this group are (in the UK) self-employed in industry generally, and those employed in agriculture and (worldwide) mobile offshore unit crews. It is thought that the better safety record of factory work is due to a combination of improved technology and effective supervision, factors which are lacking or weaker in the case of agriculture and the self-employed. For offshore mobile units the high casualty rate is ascribed to the high intrinsic risks of exploration and drilling, due for example to blowouts, explosions and fire, and to the vulnerability of offshore craft to severe weather. It is considered that improved evacuation and rescue systems could ameliorate this situation.

In one major industry, hydrocarbon processing, the incidence and

Table 1.30 Cause of total loss of units: allocation of main categories

Main category	Category according to original service document: Aircraft – CAA	Aircraft – Boeing	Hydrocarbon processing	Offshore	Shipping
Mechanical failure	Inflight fire Airframe failure Doors/windows opening Electrical systems failure Power unit failure Tyre burst	Airplane systems Maintenance	Mechanical failure	50% Explosion, fire, spillage Machinery defect Helicopter accident Structural damage	50% of Foundered Explosion
Natural hazard	Aquaplaning Bird strike Hail damage Ice/snow accumulation Lightening strike	Weather	Natural hazard	Blowout 50% of Anchor failure Capsize Collision Towing accident	50% of Foundered Lost Collision Contact Grounded
Operational error	Collision Fuel exhaustion Instruments incorrectly set Runway accident	Flight crew Airport Air traffic control	Operational error Design error Process upset	50% of Anchor failure Capsize, collision Explosion, fire, spillage Towing accident	50% of Explosion Collision Contact Grounded Lost
Sabotage	Crew shot Bomb on board Shot down	(Excluded)	Sabotage	(Not listed)	(Not listed)
Unknown, other	Loose cargo 3rd party accident	Unknown or awaiting reports	Unknown	Other	—

severity of catastrophes is increasing with time. The increasing capital and consequential losses are having an unfavourable effect on the worldwide insurance markets. The problem is associated primarily with an increase in the incidence of vapour cloud explosions due mainly to the rupture of piping. It is suggested that the increase in the quantity of gaseous hydrocarbons that is being handled in process plants has raised the intrinsic hazard of the operation, but the evidence suggests that this is not the only factor, and that other, as yet undisclosed, factors may play a part.

References

1 Warner F, 'Quality and risk', *Quality Forum* 1993 **19** 93–98.
2 *Worldwide Offshore Accident Databank* (WOAD) 1990 (published biennially), Bureau Veritas, Oslo, Norway.
3 *Health and Safety Committee Annual Report 1992–3*, HSE Books, London, 1993.
4 *Statistical Summary of Commercial Jet Aircraft Accidents. Worldwide Operations 1959–1992*, Boeing Commercial Airplane Group, Seattle, USA, 1993.
5 *Accidents to Aeroplanes*, Civil Aviation Authority, London Airport (Gatwick), 1993.
6 Peel C J and Jones A, 'Analysis of failures in aircraft structures', *Metals Mater* 1990 **6** 496–502.
7 Mahoney D, *Large Property Damage Losses in the Hydrocarbon–Chemical Industries*, Marsh and McLennan, Chicago, 15th edition, 1993.
8 Mahoney D, *Large Property Damage Losses*, Marsh and McLennan, Chicago, 14th edition and analysis, 1992.
9 Lees F B, *Loss Prevention in the Process Industries*, Butterworth, London, 1984.
10 Drogaris G, *Major Accident Reporting System*, Elsevier, Amsterdam, 1993.
11 Jenkins C, *Oil Economists Handbook*, 5th edition, Elsevier Applied Science, 1989. Updated by: Anon, *BP Statistical review of world energy*, British Petroleum, London, 1994.
12 Collins J A and Monack M L, 'Stress corrosion cracking in the chemical process industry', *Mater Protection Performance* 1973 **12** 11–15.
13 Nelson P and Still J R, 'Metallurgical failures on offshore oil production installations', *Metals Mater* 1988 **4** 559–564.
14 Lancaster J F, unpublished report.
15 Osman R M and Raziska P A, 'Correlation of reformer catalyst tube failure data', *AIChE Symposium on Ammonia Plants and Related Facilities*, Vol. 17, 1975.
16 Salot W J, 'High pressure reformer tube operating problems', *AIChE Symposium on Ammonia Plants and Related Facilities*, Vol. 17, 1975.
17 ASTM Special Technical Publication 91A.
18 Eyers J and Nisbett E G, *Proc Inst Mech Engrs* 1964/5 **179** Part II Paper 9.
19 Phillips C A G and Warwick R G, *Survey of defects in pressure vessels*, UKAEA report AHSB (S) R 162, 1968.
20 Kellermann O, *Present Views on Recurring Inspection of Reactor Pressure Vessels in the Federal Republic of Germany*, Technical Reports Series No. 81, Int. Atomic Energy Agency, Vienna, 1968.

21 Lancaster J F, 'Failures of boilers and pressure vessels; their causes and prevention', *Press Vessels Piping* 1973 **1** 155–170.
22 Bush S H, 'Pressure vessel reliability' *J Press Vess Technol* 1975 **97** Series J No 1 54–70.
23 van Rongen H J M, 'Prediction of fracture behaviour of welded steel', *Lastechniek* 1991 57 179–187.
24 Rintaman R *et al, Prevention of Catastrophic Failure of Pressure Vessels and Piping*, Technical Research Centre of Finland, Research Reports 515, 1988.
25 French D N, *Metallurgical failures in fossil fired boilers*, John Wiley, New York, 1983.
26 Port R D and Herro H M, *The Nalco Guide to Boiler Failure Analysis*, McGraw-Hill, New York, 1991.
27 *Lloyd's casualty returns; annual statistical summary*, Lloyd's Register of Shipping, London. First published 1891.

CHAPTER 2

Supercatastrophes

Introduction

In the previous chapter a catastrophe was taken to be the loss of any substantial human artefact such as a ship, an aircraft or a unit of process plant. In most instances a minimum size limit was set: 100 tons displacement for ships and 60 000 lbs weight for jet aircraft. In the case of process plant the largest dollar losses were recorded, such that the minimum loss was in the region of US $10 million. The great majority of these catastrophes did not come to the attention of the general public, however, for the simple reason that in most cases only a few lives were lost, and often there was no human loss at all. There are a few exceptions to this rule; the failure of Kings Bridge in Melbourne, Australia, is a good example. Four main girders of this bridge collapsed when it was in service, but the deck was supported by concrete structures below and there were no casualties. There was, however, considerable inconvenience and cost, and no little embarrassment to the engineers responsible when it was found that many welds in other parts of the bridge were cracked. This disaster was widely reported.

Large-scale disasters, and particularly those causing many deaths, cause great concern in the country that is affected. When a disaster happens it is normal practice for the government concerned to set up an investigation into its causes, so that recommendations can be made for its avoidance in the future. In broad terms, such enquiries have little effect on the incidence of disasters. There is no evidence that the enquiry into the *Titanic* had any effect on the rate of loss of shipping which, as has been seen in Chapter 1, was falling before the sinking and continued to fall at a similar rate subsequently. Likewise, following the investigation of the destruction of the caprolactam plant at Flixborough, UK, by a vapour cloud explosion in 1974, the incidence of vapour cloud explosions has increased at what appears to be an accelerating rate. The conclusion is hardly surprising, because enquiries into a single incident focus on the cause of that one incident, whereas catastrophes in general have a multiplicity of causes.

Nevertheless, the history of catastrophes can be very instructive. It is characteristic of many such incidents that a number of unfavourable circumstances coincide so as to magnify the damage or human loss, and turn an accident into a supercatastrophe. To know how such multiple causes interact may well improve safety procedures in a way that knowledge of the statistics alone cannot. Therefore this chapter is devoted to a description of major catastrophes, and to an attempt to draw some conclusions from their history.

The *Titanic*

Predicting the end of the world has been the favourite occupation of gloomy prophets ever since such persons came into existence. Pentti Linkola, of Saaksmake, Finland, believes this will be due to over-population, and advocates the culling of mankind. He pictures life as a lifeboat: 'Those who hate life try to pull more people on board, and drown everybody. Those who love and respect life use axes to chop off the extra hands hanging on the gunwale'.

The doomed ship has always figured largely in the literature of human disaster. William Rees-Mogg (who also reported the thoughts of Mr Pentti) put it thus:[1]

> The sinking of the Titanic in 1912 is one of the most popular metaphors for the condition of mankind. The reckless speed, the competition for the Blue Riband of the Atlantic, the wealth and luxury of the First Class, the cramped poverty of the steerage, the foolishness of those who believed the ship unsinkable, the negligence of the captain, the arrogance of mankind in the face of nature – all make the story a parable for today.

Moreover it has generated its own myths. It was suggested at the time, and is implied in the quotation given above, that the excessive speed was due to an intention to make the fastest transatlantic crossing and capture the Blue Riband. But this was a physical impossibility; the rival Cunard ships had more power and less weight. The White Star line, which owned the *Titanic*, aimed to compete in size and luxury, not in speed. A more recent theory is that there was a fire in one of the bunkers, and the captain was anxious to reach New York before the ship burst into flames. The tale has no end.

Design features

To forget the myths for the moment: the actuality began in the shipyards of Harland and Wolff in Belfast during the early years of the twentieth

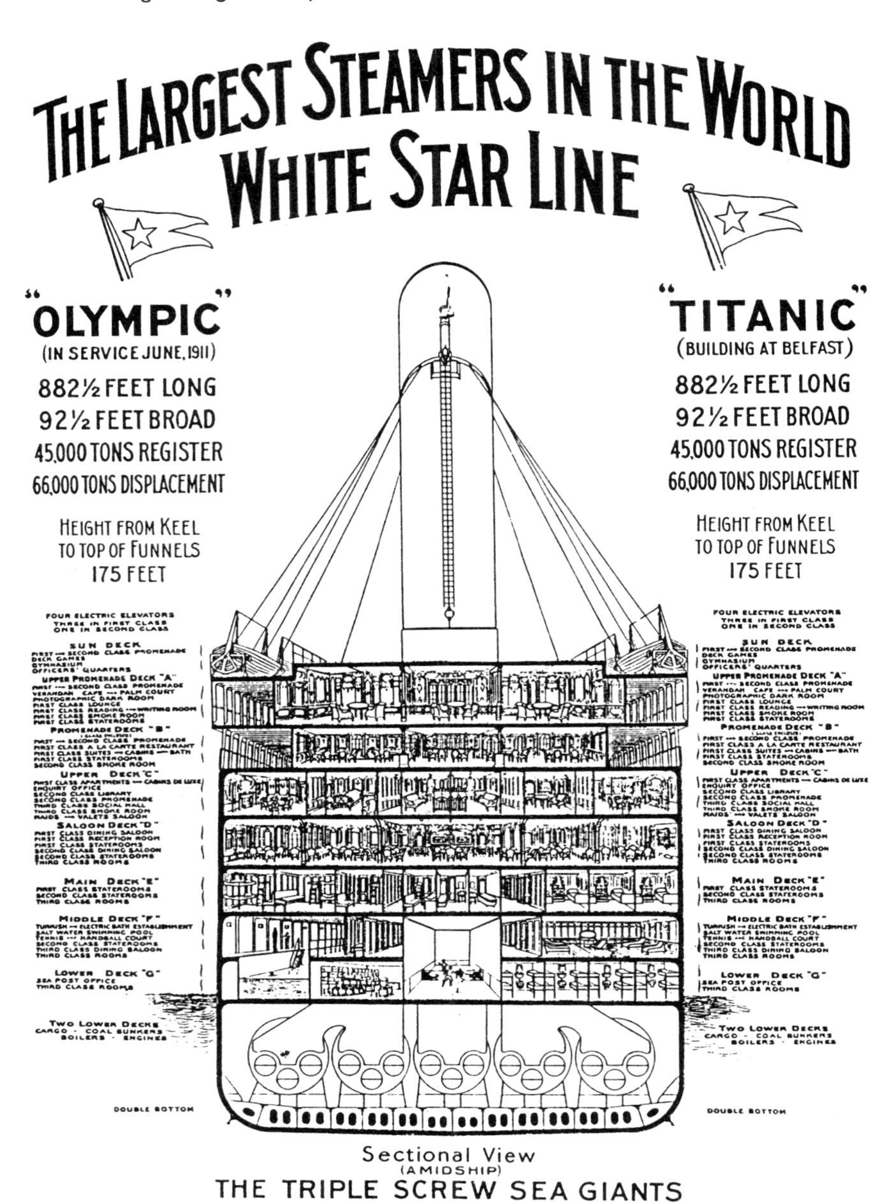

2.1 The White Star Line, based at Liverpool, was at the time when the *Titanic* sank owned by J P Morgan, the American financier and railroad millionaire.

century. The *Titanic* was one of three sister ships, the other two being the *Olympic* and the *Britannic*. They were built to the latest available technology. Originally it was intended that they should be twin-screw ships driven by reciprocating engines, but at a later stage a third screw was added, driven by a steam turbine. There was a double bottom, but the feature that made headlines was that the hull was divided into watertight compartments. At the base of each bulkhead that formed a compartment was a door, normally held up by an electromagnet. These doors were operated from the bridge; in an emergency an alarm would sound, and after an interval of time long enough to allow the crew to escape, the door came down. Those that failed to get through the doors in time would escape by ladders. The ships were designed to remain afloat with two compartments flooded. It may have been foolish to think that the *Titanic* was unsinkable, but it was in fact reasonably secure against most of the likely hazards. At the present time loss due to a grazing collision (contact) is relatively rare, amounting to about 4% of the total, and there is no reason to believe that matters were different in the days of the *Titanic*. Most other accidents would have flooded one or at most two of the compartments, so it could be argued that the ship was about 95% unsinkable.

When it was launched on 31 May 1911 the *Titanic* was the largest ship in the world, with a length of 825.5 ft and 92.5 ft across the beam. The displacement was 46 382 tons, as compared with an average (for 1910) of about 1400 tons. After being fitted out and undergoing sea trials, the *Titanic* left Southampton to start her maiden voyage on 10 April 1912 with 922 passengers on board. While she was leaving, there was a near-collision with a moored vessel. When two ships pass at close quarters the water between them flows at an increased speed, with a corresponding fall in pressure. As a result, the ships move towards each other. This happened in the case of the *Titanic*, and disaster was only averted by prompt action on the part of the master, Captain Smith, and by the accompanying tug-boats.

Later the same day she called at Cherbourg and took on 274 more passengers. On the following day she stopped at the Irish port of Queenstown and took on a further 120 people, mostly Irish emigrants to the United States. The *Titanic* finally set off across the Atlantic on 11 April with a complement of 2228 passengers and crew. The UK Board of Trade regulations required that ships with a displacement of over 10 000 tons should carry 16 lifeboats. The *Titanic* carried, in addition to the required number, four collapsible boats, giving a total capacity of 1178 persons, about half the number on board, and about one-third of the maximum complement of 2300. By contrast, regulations at that time

2.2 The *Titanic* at berth just prior to its fateful voyage.

2.3 '. . .the wealth and luxury of the First Class. . .': the grand staircase of the sister ship *Olympic.* That of the *Titanic* was similar, and it may be seen today in ghostly video pictures taken inside the wreck.

specified that cargo ships should carry lifeboats with a capacity of twice the ship's complement, it being argued that when a ship was sinking, it could well list in such a way that only half the lifeboats could be launched.

The sinking

On the night of 14/15 April the ship was steaming at full speed, about 22 knots. The weather was still and frosty, cloudless but with no moon, and the sea dead calm. There had been warnings of ice ahead by wireless from other ships. The most recent call was from the *Californian*, a British ship located a few miles away, and stationary because of field ice in the vicinity. At the same time the radio operator of the *Titanic* was passing messages from the passengers to US destinations and the backlog was such that he declined to accept the call. Shortly after this the radio operator on the *Californian* finished his watch and went below. A short time later still, the *Titanic* struck an iceberg and sank.

2.4 Londoners outside Oceanic House in Cockspur Street near Trafalgar Square read about the tragedy.

It was by no means unusual to encounter icebergs in the North Atlantic in April, but in 1912 the ice had drifted further south than usual. The normal precaution against colliding with an iceberg at night was to extinguish or mask lights forward of the bridge and post lookouts. This was done on the *Titanic*, with two men aloft and two officers on the bridge. It was customary to keep watch by naked eye; binoculars or telescopes restrict the field of view. All should have been well, but when the iceberg was sighted it was only about a quarter of a mile ahead. The officer of the watch immediately ordered full astern and put the helm hard over. The distance was too short, however, and ship received a series of glancing blows over a length of about 250 ft. These caused plating to buckle and rivets to shear or fail in tension, cracking the caulking. Water then flooded into five, possibly six, of the forward compartments.

The impact occurred at 11.40 pm and some time elapsed before the captain gave the order to man the lifeboats. Women and children were to board first, but it proved difficult to persuade married women to leave their husbands behind. In the steerage particularly, passengers were reluctant to leave their belongings (mostly the only things they possessed), particularly as the stern portion of the ship was dry. Eventually the 16 boats and 1 collapsible were launched and other collapsibles pitched into the sea.

Throughout this period the crew behaved in exemplary fashion, attending to their duties to the end. The engine-room staff maintained power and the lights shone brilliantly until just before the final plunge; none of these men survived. Only one of the senior officers was saved; this was Second Officer C H Lightoller. He had dived in after freeing the last of the collapsible boats, and after being pulled down he was blown to the surface by a sudden uprush of water, and then managed to scramble on to an upturned collapsible. Captain Smith remained on deck directing operations until it was clear that no more could be done; he then walked forward calmly to take his place on the bridge. The ship went down at 2.20 am, with the loss of over 1500 lives.

The rescue

Several ships picked up the wireless distress signal CQD, the nearest of which was the *Carpathia*, about 30 miles distant. All these ships steamed at full speed towards the wreck until the *Carpathia* signalled her arrival. Some put an extra watch in the stokehold in order to find a little more speed. The *Carpathia* picked up the survivors. According to the master of the *Carpathia* these numbered 705, but the purser later added six more making a total of 711. This figure was adopted officially, but the captain subsequently maintained that his original number was correct.

The enigmatic feature of this period is the role of the *Californian*. There was no radio operator on duty, so the distress calls on the wireless were not heard. As well as radio signals, the crew of the *Titanic* also indicated their situation by sending up distress rockets. Officers on board the *Californian* saw the flashes from these rockets, but were uncertain about their significance. They tried to communicate with the unknown vessel by Morse lamp but without success. After a time the ship disappeared and was presumed to have sailed off. These events were reported to the master, Captain Lord, who was resting in his cabin. Then when the morning watch came on it was decided to rouse the wireless operator. They then learnt that the *Titanic* had foundered and at 6 am steamed towards the scene of the disaster, arriving in the vicinity of the *Carpathia* at 8.30 am.

The aftermath

The *Carpathia* sailed to New York and her arrival was the occasion of much public excitement and press speculation. The US Senate almost immediately set up its own committee of inquiry under the chairmanship of Senator William Alden Smith of Michigan. Smith's knowledge of ships and seafaring matters was exceedingly small, but this did not prevent him coming to positive conclusions about the cause of the tragedy: it was due primarily to overconfidence and indifference to danger on the part of Captain Smith, whilst Captain Lord of the *Californian* bore a heavy responsibility. All aspects of the escape were criticised, and only Captain Rostron of the *Carpathia* received any praise. The Senator did not improve Anglo-American relations.

The British formal investigation was led by a lawyer, Lord Mersea. It exonerated Captain Smith on the grounds that he was following normal practice, but said that had Captain Lord acted properly, many more lives could have been saved. Controversy over both these conclusions continues to the present day.

The officers and crew of the *Carpathia* were much praised for their rescue effort, and several of their number received medals or decorations. The *Olympia* went into dock shortly afterwards and emerged with the double bottom extended up the sides of the hull, and with rows of shining lifeboats along its decks. The Board of Trade regulations were amended, and lifeboat drill became a regular feature of the transatlantic (and other) crossings.

Finding the wreck

The development of techniques for underwater exploration began in earnest in the 1930s, and one of the pioneers was Dr Piccard, who used a spherical diving shell known as the Bathyscape. This was the start of a long tradition of French underwater work. The US navy began to take a serious interest after the loss of the submarines *Theseus* in 1963 and *Scorpion* in 1968. The needs of the offshore oil and gas industry for exploration inspection and maintenance gave further impetus to this development, leading to the current generation of submarines, which are mainly unmanned vehicles controlled from surface ships. Garzke *et al*[2] gave an excellent review of this development, starting with Alexander the Great, who is said to have descended into the waters of the Aegean sea in a glass bell in order to view the wonders of undersea life.

It is appropriate that the wreck of the *Titanic* was eventually located by a collaborative effort by French and US explorers. The French team first narrowed the field of search using a sonar scan, then Dr Ballard and his colleagues from Woods Hole Oceanographic ·Institution employed a remotely controlled camera sled *Argo* to examine the remaining area visually[3], and in September 1985 finally located the wreck at a depth of 1200 ft.

It came as a surprise that the remains of the ship were in two large pieces, 1970 ft apart. It had been generally supposed (although some eyewitnesses had it otherwise) that after colliding with the iceberg the ship sank substantially intact. This, indeed, was the conclusion of the Board of Inquiry. The distance between the two parts is about one-sixth of the water depth, suggesting that the separation took place some way below the surface. The bow portion is more or less intact, but the stern has been crushed into a tangled mass of wreckage.

Garzke *et al*[2] suggest the following sequence of events:

1 Just before the final plunge, the stern was raised clear of the water at an angle of 45–60 degrees. This induced high stresses in the deck plating and girders, which were enhanced by the presence of the two heavy reciprocating engines.
2 The initial break probably started in 'B' deck between the compass platform and the third funnel. About this time the boilers in No 1 Boiler Room came loose and were forced up against the deck girders, allowing the sides to compress inwards.
3 The ship sank when only partially severed, but at some distance below the surface a final separation took place.
4 The forward portion was already flooded and pressures were equalised, so it did not suffer any significant damage due to water pressure. The stern portion on the other hand still contained air and crushed as it descended.

This proposed sequence makes sense, and conforms with the majority of eyewitness accounts.

The cause of the tragedy

The first question to answer is: how is it possible that four alert men were unable to see a major obstacle until it was too late to avoid it? They were experienced seamen, accustomed to the job of spotting ice, so it has to be concluded that the iceberg was not visible except at relatively close quarters. One of the lookouts in the crow's nest, Fleet, described it as a 'dark shadow on the water', and the other man, Lee, told the Mersea enquiry 'It was a dark mass that came through the haze and there was no white appearing until it was just close alongside the ship, and that was just a fringe at the top'.

Sir Ernest Shackleton, the polar explorer, explained this phenomenon to the enquiry. Exceptionally, an iceberg may melt or disintegrate in such a way that it becomes top heavy and capsizes. The underside which is then exposed appears black owing to contamination with earthy matter and porosity. He had twice seen such icebergs in the North Atlantic.

The other factor making for invisibility was the dead calm: there was not even a swell. Consequently there was no ring of breakers around the berg. Lightoller suggested that it might be another 100 years before these two conditions occurred together again. He was almost certainly right because there had not been any sinking due to an iceberg on the transatlantic shipping lanes up to that time, and there has not been another to this day. One ship, the *Hans Hedtoft* struck an iceberg in 1959 off the coast of Greenland and went down with the loss of all 95 passengers and crew. So although sinking through collision with an iceberg is a rare event, it could still happen even with the navigational aids available in the late 1950s.

As to whether the speed was too high: well, with hindsight, it was indeed so. But by prior standards it was quite normal to proceed at full speed through the ice provided that visibility was good. Experience had indicated that this practice was safe provided that experienced lookouts were posted. Lord Mersea's assessment of the situation would seem to be a reasonable one.

As to the *Californian*: there is a fair body of opinion nowadays that would reject the charge that the Captain and crew were wilfully negligent. After all, they did steam at full speed through the ice to the disaster area on the following morning. And it did take over two hours to get there. So the notion that she could have averted the tragedy is open to question.

In fact it would seem, if the above analysis is correct, that the *Titanic* suffered the classic type of catastrophe where several unfavourable circumstances or events, each one of which has a low probability, coincide. The possibility that the ship would founder on its first voyage would have appeared to the passengers who set off from Southampton to be a virtual impossibility; and for good reason, for that is exactly what it was.

The brittle fracture theory

Following the discovery of the wreck, samples of the steel plate used in construction have been obtained and tested. The results showed that the notch-ductility of the material at 31°F (the water temperature at the time of the sinking) was low; in other words the steel was notch-brittle. This fact led Garzke *et al*[2] to suggest that brittle behaviour of the plate could have 'contributed to the hypothesized rivet or plate failures'. This rather modest suggestion has been much improved in press reports, for example in *The Times* of 17 September 1993:[4]

> Maritime experts who presented their findings yesterday to the Society of Naval Architects and Marine Engineers in New York, blamed the rapid flooding on extensive cracks in the steel plates .... The team attributed the cracking to 'brittle fracture' in which low-grade steel breaks violently when cold, rather than bending ... better quality steel might have kept her afloat for another two hours.

In fact there is not the slightest evidence that brittle cracking had anything to do with the *Titanic* disaster. When steel is in a notch-brittle condition, a plate containing a crack and subject to a tensile stress may fail suddenly at a relatively low value of stress. However, if an uncracked specimen of the same steel is subject to a normal tensile test at the same temperature it will fail in a normal ductile manner, although it will stretch plastically somewhat less than a more notch-ductile steel. Likewise, when subject to a bending load, it will, in the absence of a notch or crack, bend and not break. It is only when there is a notch or crack of sufficient length already present in the material that it will behave in a brittle way.

The wreck itself provided clear evidence of ductile behaviour. The stern portion was severely damaged by implosion, but did not break up. The forward portion, apart from local areas, is bent but not broken. And according to most witnesses, the ship went down in one piece, suggesting that the bottom plates must have bent initially, and then parted after she was below the surface.

The difficulty, indeed, is to explain why she remained afloat so long after such a severe impact. The Harland and Wolff designer, Edward Wilding, calculated that the area of leakage was 144 square inches. The length of the damage, as noted previously, was reckoned to be about 250 ft. If the gap had been uniform along this distance it would have been about 3/64 inches or 1.2 mm wide. Suppose that there were six gaps each 2 ft in length, then the average width would be 1 in. These figures are consistent with a model in which the plating was buckled sufficiently to break the rivets and displace the caulking. In such a case the rivets would have failed or would have been torn out regardless of their ductility. The real problem seems to have been the length of the damaging contact, which meant that five or six of the forward compartments were flooded.

The report in the London *Times* newspaper implied that a better grade of steel could have been used to build the ship, and this could have saved lives by keeping the ship afloat for another two hours. In fact, there was at the time only one grade of shipbuilding steel. The required properties of this grade were specified by Lloyd's Register of Shipping, which also inspected and approved the steelworks and witnessed the testing of samples from each melt. The suggestion that inferior material was used is entirely without foundation.

Riveted ships do not suffer from catastrophic brittle failure. The brittle fracture problem arose some 30 years after the building of the *Titanic*, when the first all-welded ships went into use. With a monolithic hull, a running crack could split the ship in two; in a riveted ship the crack only got as far as the edge of the plate in which it initiated. Such localised cracking did occur from time to time, but would always require an initiating defect, usually a fatigue crack. Fatigue cracks take some time to develop. The *Titanic* was a new ship, so even localised brittle fractures would have been highly improbable. It may be concluded that although the hull plates of the *Titanic* were in a notch-brittle condition at the time of her encounter with the iceberg, they were very unlikely to have failed in a brittle fashion; all the evidence suggests that they bent, and did not break.

Ferry disasters

Ferry accidents can be particularly distressing because the victims are often ordinary people going on holiday, visiting friends or relatives or simply enjoying a cruise without any thought of dire consequences. The loss of over 900 lives including those of pensioners and children when the *Estonia* capsized and sank in the Baltic sea in September 1994 was particularly tragic.

So far as loss of life is concerned, the world's worst shipping accident occurred when the Philippine ferry-boat *Dora Paz* was in collision with an oil tanker. In the ensuing fire nearly 4400 people were killed. However, safety on ferry-boats in the Philippines is very far from that in the developed countries. The average annual death toll for the period 1980–89 was 620. In the UK, a country with a similar population, the average death rate of passengers in ferry operations from 1950 to 1988 was 7 per year.[5] Evidently the level of risk in these two areas is very different, and they need to be treated separately. The comparison will have a special interest. As noted elsewhere, the statistics quoted in this book relate primarily to industrialised countries; the Philippine ferries (of which J Spouge has provided an excellent account[6]) are an interesting example of such operations in a country which has yet to develop an industrial base and the full regulatory activities that are necessary for human safety in industry and transport.

European ferries

The Estonia disaster

At 7 pm on Tuesday, 27 September 1994 the motorship *Estonia* left the port of Tallin in Estonia bound for Stockholm. The *Estonia* was a roll-on, roll-off (ro-ro) ferry which had a capacity of 460 cars and space for 2000 passengers. She was equipped with both bow and stern doors, the bow door being of the visor type; that is to say it hinged upwards to open. The vessel itself had been built in Papenburg, Germany, in 1980, and was constructed in accordance with the best contemporary practice.

At about 1.20 am on the morning of 28 September the third engineer Margus Treu, who was in the engine room at the time, heard two or three strong blows that shook the whole ship. Shortly afterwards the vessel heeled suddenly, throwing passengers off their feet. For a time the list stabilised, such that some of the passengers had time to dress and to climb up on deck. Then, about 15 minutes later, the ship turned on its side and sank.

There was no time for an orderly evacuation, and in any event power was lost when the ship first heeled, so lifeboats could not have been lowered. A number of liferafts were flung into the sea, and these were the means by which some of the passengers survived. The weather was squally with waves up to 30 ft, and in several cases individuals were washed off their raft and had to fight their way back. The first survivors were picked up by the ferry-boats *Mariella* and *Symphony*, which were nearby. Then at first light helicopters took off from the mainland to help in the rescue operation. Altogether 140 were saved, most being young men. Over 900 lives were lost, making this the worse peacetime ferry disaster in European waters.

The vessel lay at a depth of 230 ft near the island of Uto, off the coast of Finland. As soon as the sea became calmer, video cameras mounted on underwater craft were sent down. In addition to the outer door, ferries are required by the classification societies to have an inner door, which is held in place by hydraulic clamps and which forms the main barrier against the ingress of water. The video pictures were clear, and showed that the outer door of the *Estonia* had been torn off and was missing, whilst the inner door had been forced open such that there was a gap of about 3 ft along its upper edge. A later search revealed the missing door on the sea-bed at some distance from the ship.

This account is based primarily on reports in *The Times* newspaper. The results of further investigations and enquiries will no doubt appear in due course.

The safety of roll-on, roll-off ferries

The loss of the *Estonia* confirmed the worst predictions of those who believed that this type of ship was, because of the design, unstable and liable to capsize rapidly and catastrophically if water invaded the main vehicle deck. This loss occurred not very long after the International Maritime Organisation (an agency of the United Nations) had agreed that following a collision, ferries should be capable of remaining afloat and upright for at least 45 minutes, leaving sufficient time (in theory, at least) for the passengers to escape. The agreement failed to obtain full international support, nor, as will be seen later, would it be realistic to expect this to be so. Nevertheless some countries have adopted it as an objective. The *Estonia* disaster has shown that there is much work to be done before this objective is achieved.

Roll-on, roll-off ferries developed during the post-1945 period from the vessels that were used for the Normandy and other sea-borne invasions by Allied forces. Indeed some of the early ferries used adapted LSTs (Landing Ship, Tank). They are characterised by shallow draught (necessary to serve ferry ports), a vehicle deck which runs the length of the ship and is normally open, and access doors with a low freeboard to give a level or gently sloped loading ramp. Early versions had side doors, but these cause difficulties in loading and unloading, so that most of the contemporary designs have bow and stern doors. There are two types of door, the visor and the clam types. The visor, as tragically demonstrated in the case of the *Estonia*, is less safe because wave motion tends to open it. The clam type is flat-faced and closes by hinging upwards. It is more difficult to make watertight, and is less elegant, but is less subject to weather damage. Neither type is expected to be completely watertight, and the water seal is provided by an inner door, as noted earlier.

In the larger boats it is possible for vehicles to turn through 180° and the bow door can be eliminated. This greatly reduces the risk, but at the expense of speed in turn-around. For the future it is likely that vessels operating in rough seas such as the Baltic and North Sea will have a stern door only. New vessels for North Sea operations have been so constructed, and the replacement for the *Estonia*, the *Veronia*, is to have its bow doors welded up.

Ro-ro ferry ships may become unstable for a number of reasons, including shallow draught, low freeboard and the long vehicle deck. Flooding of the vehicle deck is the most-feared event, because it can result in a very rapid capsize, such that no proper evacuation is possible. When a substantial volume of water enters the car deck it flows either to one side or the other, causing a rapid heel. After the initial movement the vessel will stabilise temporarily, but will then continue to heel and finally capsize or, in shallow water, settle on its side.

This type of flooding is a special feature of ro-ro ferries but is only one of the many accidents to which these vessels may be subject, as will be seen below.

The record of ro-ro accidents

Since World War II, ferries in industrial countries have experienced seven major incidents resulting in the sinking of a ro-ro vessel, and these are listed, together with a brief summary of the causes, sequence of events and casualty rate, in Table 2.1. Two of the seven were due to flooding through the bow door. One of these, the *Herald of Free Enterprise* (Fig. 2.5), was a loss caused by gross maloperation, and the procedures of the operating company concerned have been modified so that a repetition of such an accident with its ships is highly improbable. Such procedures do not, however, offer any protection against the *Estonia* type of accident, and it remains to be determined how this problem may be resolved. Of the other sinkings, one was due to failure of the stern door, and one to side doors being stoved in; in other words the bow door is not the only vulnerable point. One sinking was due to a collision, but apart from that of the *Herald of Free Enterprise* the remainder were all primarily caused by severe weather or heavy seas. In all cases the ships finally capsized and sank, but except for the *Estonia* and the *Herald of Free Enterprise* where there was a massive inundation of the vehicle deck, they all remained afloat long enough to organise an evacuation and, in some cases, to launch the lifeboats. Thus, most ferry sinkings (about 70%) were due to normal shipping-type hazards. On the other hand, 85% (1103 out of 1295) of the deaths, as presently recorded, were due to flooding of the vehicle deck. For European ferries this remains the most important problem.

Table 2.1 Major European ferry sinkings (in part from Ref. 5)

Date	Ship	Location	Cause of flooding	Where flooded	Sequence of events	Number on board	Dead
1953	*Princess Victoria*	Irish Sea	Large wave burst open stern door in rough seas	Car deck and starboard engine room	Listed, reaching 45° in 4 h. Capsized and sank after 5 h. Ship abandoned	172	134
1966	*Skagerak*	Skagerak	Heavy seas stove in side doors	Engine room	Sudden heel, cargo shifted. Ship abandoned except for 11 crew. Capsized and sank when under tow	144	1
1968	*Wahine*	Wellington, New Zealand	Grounded on rocks in severe weather	Bow and stern below decks	Remained afloat for 5½ h, then started to heel. Capsized and sank after 7½ h. Abandoned	735	51
1980	*Zenobia*	Cyprus	Failure of auto-pilot, causing list	General	Passengers and crew evacuated. Vessel capsized and sank during attempted salvage	151	0
1982	*European Gateway*	Off Felixstowe, UK	Rammed by ro-ro ferry *Speedlink Vanguard*	Engine room	Heeled to 40° in 3 min at which point bilge grounded and she rolled on her side in 10–20 min	70	6
1987	*Herald of Free Enterprise*	Zeebrugge	Bow doors not closed prior to sailing	Vehicle deck	Bow wave high enough to flood lower vehicle deck. Heeled to 30°, then paused for few seconds, finally grounding on her side	539	193

1994	*Estonia*	Off coast of Finland	Outer bow door torn off in heavy seas, inner door forced open	Vehicle deck	Vessel heeled suddenly, stabilised for few minutes, capsized and sank in 15–20 min	1050[a]	910[a]

[a]Approximate figures, based on newspaper reports.

2.5 The *Herald of Free Enterprise* capsized outside Zeebrugge harbour.

The numerical risk

Spouge[5] calculated the risk of passenger travel on ferries originating in the UK in 1988. At that time there had been a total of 333 deaths on such ferries and, assuming a start date of 1950, the annual rate was 9, comprising 7 passengers and 2 crew. The total number of ferry crossings annually was 28 million, so that a passenger taking one return trip per year had an annual mortality risk of 5×10^{-7}. This could be compared with the annual fatality rate due to road accidents, which in 1988 was 1×10^{-4}. The fatality risk for a passenger taking two flights per year in commercial jet aircraft during the early 1990s was also 1×10^{-4}. In other words, the risk of taking a ro-ro ferry from the UK is very small indeed, and is nearly three orders of magnitude lower than other commonly accepted transport risks.

There have been no major incidents affecting UK ferries since 1988, so that the current (1995) risk would be marginally lower. For European ferries as a whole the calculated risk will have increased significantly due to the loss of the *Estonia*, but it is still estimated to be very low, of the order of 1×10^{-6}.

Philippines ferries

The Republic of Philippines is a country consisting of an archipelago of 7000 islands, of which 880 or so are inhabited. Most of the population lives on the eight largest islands. Following independence in 1946 there was a succession of presidents but after the election of President Marcos in 1965 a period of rapid economic development occurred. Insurgency in the north and Muslim separatism in the south led to increasingly despotic rule by Marcos and the economy stagnated. After the fall of Marcos and the restoration of democracy there was some recovery. However the country remains very poor, with a (theoretical) daily minimum wage (in 1990) of 114 pesos, equivalent to about US$5.50. The main economic activity centres around agriculture, forestry and fishing, and more recently the mining of metallic ores. The population is about 60 million and because of the lack of industrial development, many seek work abroad, as seamen, hotel staff and servants.

Poverty also means that travel between the islands is almost entirely by ferry; only a small number of people can afford to go by air. Ferry-boats are of all sizes. The smaller islands, where port facilities are lacking, are served by outrigger canoes capable of carrying up to 50 people, and by other smaller craft. The rest of the fleet consists mainly of secondhand vessels (such as that shown in Fig. 2.6), often purchased in Japan and with an average age of about 20 years. There are a number of ro-ro vessels, but outside the main cities there are few cars and the vehicle decks are often used for cargo and passengers. Passenger accommodation is spartan and third class ticket passengers on overnight trips are crowded in bunks on the open deck. A deluxe third class ticket rates a bunk under cover, but Mr Spouge warns that at night the doors are chained shut to prevent an

2.6 *Sweet Heart*, built 1965, registered tonnage 475, capacity 400 passengers.

insurge of the ordinary third class passengers, so if the ship capsizes in a typhoon there is no escape.

The accident records

Statistics for fatalities on Philippines ferries are derived mainly from newspaper reports; there are no systematic records as yet. Only the major disasters are likely to have been recorded, so that any estimates formed in this way must be underestimates. Table 2.2 lists the information available for the period 1980–89. The number of fatalities listed in this table totals 6224 so that the annual loss is over 600 persons. Spouge[6] estimates 17 million crossings per year for Philippines ferries in the late 1980s, so that the fatality risk was 3.8×10^{-5} per crossing, as compared with 2.5×10^{-7}, as calculated for UK ferries.

The Dona Paz disaster

The *Dona Paz* was a three-deck passenger ferry-boat of 2324 registered tonnage and with an authorised passenger capacity of 1518. It was built in Japan in 1963 and went into service in the Philippines in 1975, at which time it lost classification. In 1979 it was gutted by fire but was reconstructed and returned to service. On 20 December 1987 it was crossing from Tacloban to Manila, loaded with over 4000 passengers returning for the Christmas holiday. During the night it collided with the tanker *Vector*. Both vessels were engulfed in fire and sank, with the loss of (it is estimated) 4376 lives.

The tanker carried 1130 tonnes of gasoline, kerosene and diesel fuels. It was operating without a licence, without a lookout and without a properly qualified master. Two of the 13 crew managed to escape and were picked up by another ferry. On the *Dona Paz* most of the passengers were trapped in the burning ship; only 24 were able to get away and were rescued by the same ferry. The *Dona Paz* had no radio, and it was 16 hours before a rescue operation was mounted. By this time it was far too late.

Comparisons

Such disasters, and the generally high risk of ferry travel, are accepted in the Philippines because of the generally high accident rate amongst the population. Large numbers of outriggers, pump boats and motor launches that also operate as ferries suffer accidents and it has been estimated that the total loss of life at sea in the Philippines is between 20 000 and 40 000

Table 2.2 Ferry accidents in Philippines, 1980–89[6]

Date	Ship	Tonnage	Cause of loss	Number on board	Dead
April 1980	*Don Juan*	2300	Collision, sank	Over 1000	Over 121
July 1981	*Juan*	1530	Fire	Over 458	Over 57
Sept 1981	*Sweet Trip* (ro-ro)	500	Sank	Not known	1
June 1982	*Queen Helen*	500	Sabotage, explosion, fire	484	48
March 1983	*Sweet Name* (ro-ro)	580	Collision, explosion, fire	400	27
May 1983	*Dona Florentina*	2100	Fire, beached	884	0
Nov 1983	*Dona Cassandra*	487	Capsized in storm	396	177
Nov 1983	*Santo Nino*		Capsized		12
Jan 1984	*Nashra*		Steering gear failure, capsized		52
Jan 1984	*Asia Singapore*	720	Capsized in port during storm	568	21
Oct 1984	*Venus*	746	Sank in storm	351	137
Dec 1985	*Asuncion*	141	Sank	About 200	136
April 1986	*Dona Josephina* (ro-ro)	1000	Flooded, sank	414	194
Sept 1987	*Kolambugan*	770	Fire, sank	150	0
Dec 1987	*Dona Paz*	2324	Collision, fire, sank	4400	4376
April 1988	*Balangiga*		Sank in typhoon	148	63
Oct 1988	*Dona Marilyn*	2855	Sank in typhoon	481	284
Dec 1988	*Rosalia*		Sank in storm	410	400
Dec. 1988	*RCJ*		Sank in storm	53	51
Jan 1989	*Jem III*		Capsized	Over 190	Over 60
Nov 1989	*Jalmaida*		Sank in storm	Over 180	Over 7

per year.[7] By comparison, the losses on the large boats are small. Part of the problem is that the area is affected by typhoons, but in the main it is due to unseaworthy boats with unreliable engines carrying too many passengers.[6]

Such conditions, and fatality rates, apply to other poor countries. A process upset in a chemical plant located at Bhopal, India, resulted in the emission of poisonous gases. The number of deaths resulting from the accident was 3031. The reason for the excessive loss of life was that people were camping around the perimeter of the plant hoping for work or some other benefit. Life is cheap in such countries; literally so, because the Indian government offered the equivalent of about US$800 compensation for each life lost at Bhopal, whilst the owners of the *Dona Paz* paid just over US$900 to relatives of those who died in that disaster. These figures may be compared with the 1974 Athens Convention compensation limit of about US$60 000 per passenger, which itself is low by current standards.

From every point of view, therefore, it is necessary to treat statistics for accidents in third world countries quite separately from those for Europe and North America. The most important problem for European ferries is a technical one; how to improve the safety of roll-on, roll-off vessels. This problem is wholly irrelevant in the Philippines. Moreover, if all the figures are lumped together, ro-ro sinkings account for only a small proportion of fatalities; treated separately, the opposite is the case. Figure 2.7, which is a frequency–consequence plot comparing UK and Philippines ferry losses, underlines the need for separate treatment.

It appears that a similar state of affairs applies, although to a lesser degree, in the case of air transport. Harris,[8] quoting from an article in the *Guardian* newspaper, gives figures for the rate of loss of aircraft, expressed in numbers per million departures. These indicate a significantly higher loss rate in the less-developed countries. However, in this field the overwhelming majority of flights take place from and within developed countries, so worldwide figures are unlikely to be affected to any significant extent by such differences.

The *Alexander L Kielland* accident

The floating objects that have been used for offshore oil and gas production started as modified barges or ships, but have progressively departed from the norm, and some of the more recent developments look very unlike a conventional ship. The *Alexander L Kielland* was a semi-submersible drilling platform that had been adapted for use as an accommodation platform serving various individual production units in the Norwegian section of the North Sea.

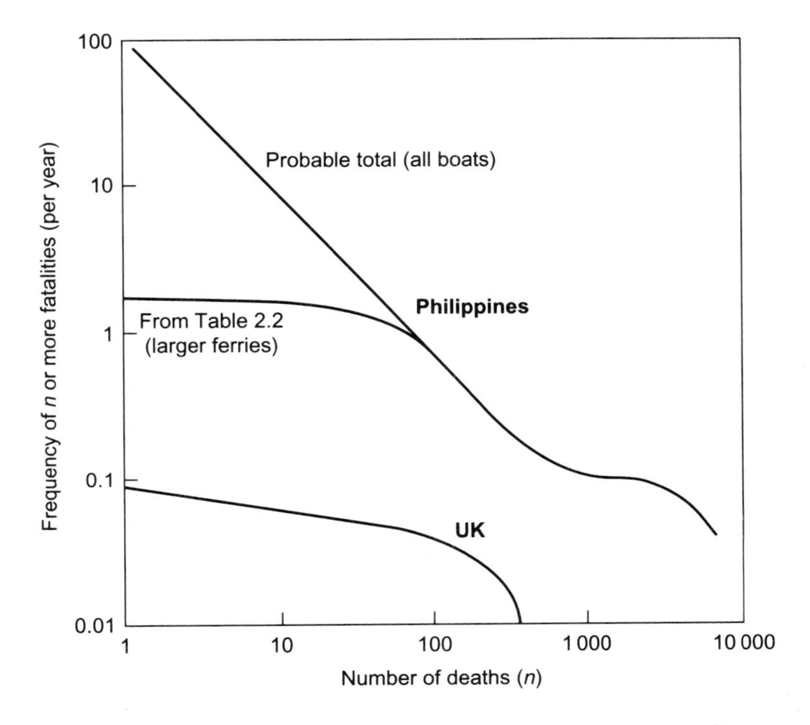

2.7 Frequency–consequence curve for ferry casualties, comparing those in the Philippines with those sailing from British ports.

Semi-submersible drilling rigs originated in the USA during the early 1960s. The object of this development was to provide more stable conditions than could be obtained in a drill ship, such that operations could be continued under more severe weather conditions. They consist of three or more submerged pontoons supporting a structure on which is mounted the operating platform and drilling equipment. It will be self-evident that a rig that obtains its buoyancy from submerged pontoons will be much less subject to wave motion than a surface ship. The first semi-submersible was *Blue Water I* which was delivered in 1962, followed by *Ocean Driller* in 1963 and *Ocean Queen* in 1965. Initially these developments proceeded empirically. However in 1978 the American Bureau of Shipping published a set of rules for the construction of mobile offshore units, followed by Det Norske Veritas and Lloyd's Register of Shipping, which published general guidelines in 1970 and 1972 respectively. Prior to the *Alexander L Kielland* disaster the record for semi-submersibles had been good. Two had been lost; *Transocean 3*, which capsized and sank after a structural failure in 1974, and *Seaco 135-C*, which was destroyed by fire following a blowout in January 1980.

The Pentagone rigs

In 1963 the Institut Français du Petrole signed a cooperative agreement with Neptune, an oil exploration subsidiary of the Schlumberger group, to develop a design for a five-pontoon semi-submersible drilling rig. This led to the construction of the first Pentagone rig, P 81, which was delivered to Neptune in June 1969. In 1970 this company, together with various other organisations, reviewed the original design and came up with a number of modifications. These developments were incorporated in a new rig, P 82, which was constructed in Brownsville, Texas. This formed the basis for a further nine Pentagones, six of which were constructed in France and three in Finland. The *Alexander L Kielland* was the seventh of this group, numbered P 89 and was delivered in July 1976. The general layout of these rigs is shown (complete with derrick) in Fig. 2.8, and Fig. 2.9 is a sketch of the deck in plan view in relation to the location of the columns. This diagram also indicates the location of lifeboats and other rescue equipment. The various structural members were dimensioned to withstand the static loading associated with a maximum wave height of 30 m. There was no attempt to design against fatigue loading. At the Commission of Enquiry hearings, the designers claimed that this mode of fracture had been taken into account in the design of the structural details and the selection of geometric form and quality of materials.

The rig was required to be towed from one location to the other by tugs, the bow of the vessel being column C and the stern lying between columns A and E. Once in position it was fixed by ten anchors, two being attached to each column by wire hawsers having a diameter of 2.75 in and a breaking load of 310 tonnes. The position could be altered by operating winches mounted on the top of the columns. Screw propellors were fitted to each column, and by operating these it was possible to modify the tension in the anchor lines.

The rig was originally developed with accommodation and rescue equipment for 80 persons. These facilities were expanded stepwise over a period of time by the addition of extra modules so that by March/April 1978 there were sleeping quarters for 318 men, with a mess hall (made by building several modules together), and two cinemas. The rescue equipment consisted of 7 lifeboats, 16 rafts under davits (launchable rafts) and 12 throw-overboard rafts. The lifeboats could hold 50 persons and the liferafts 20 each, giving a theoretical availability of 910. This was

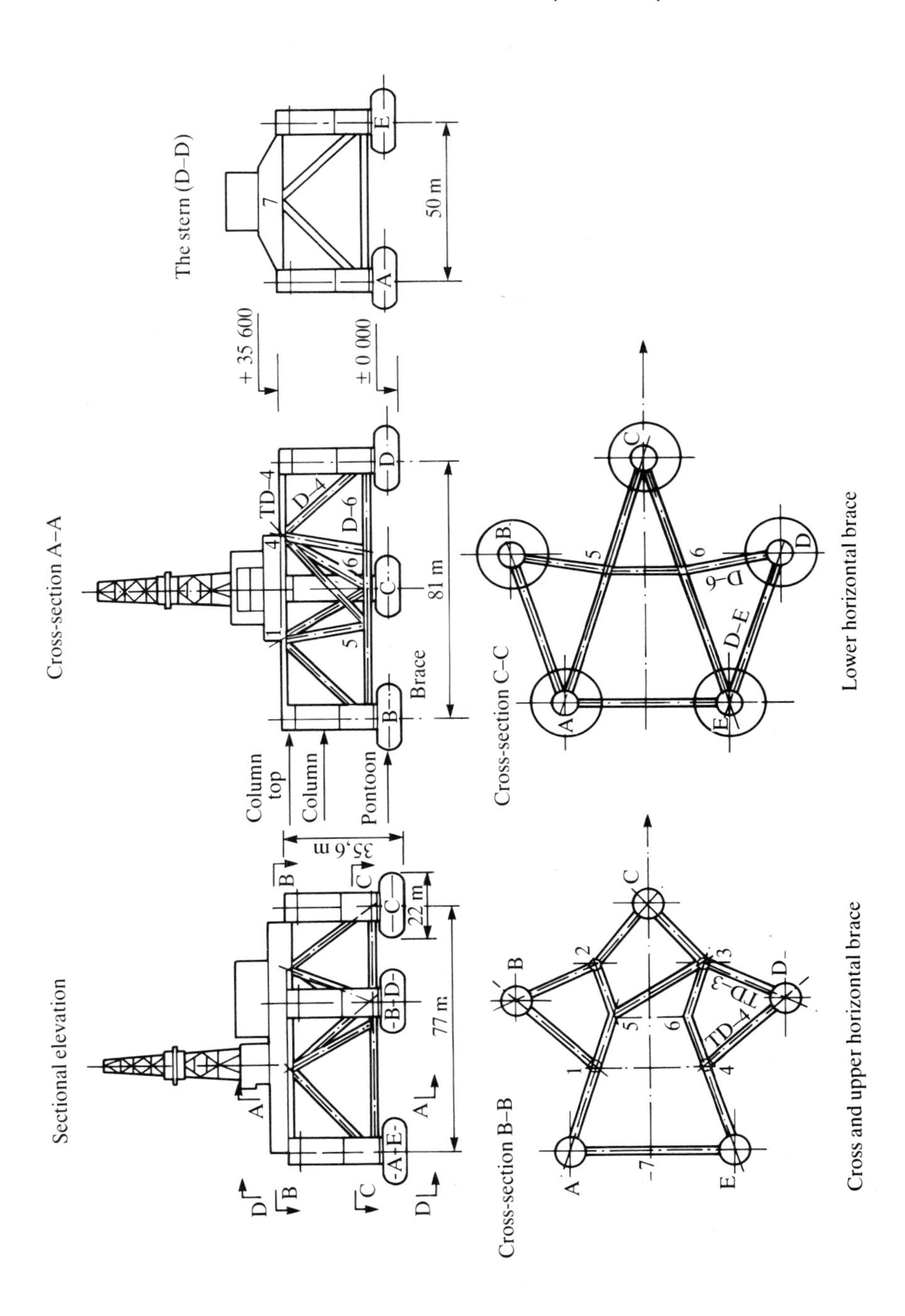

2.8 Plan and section of a Pentagone type semi-submersible rig.[9]

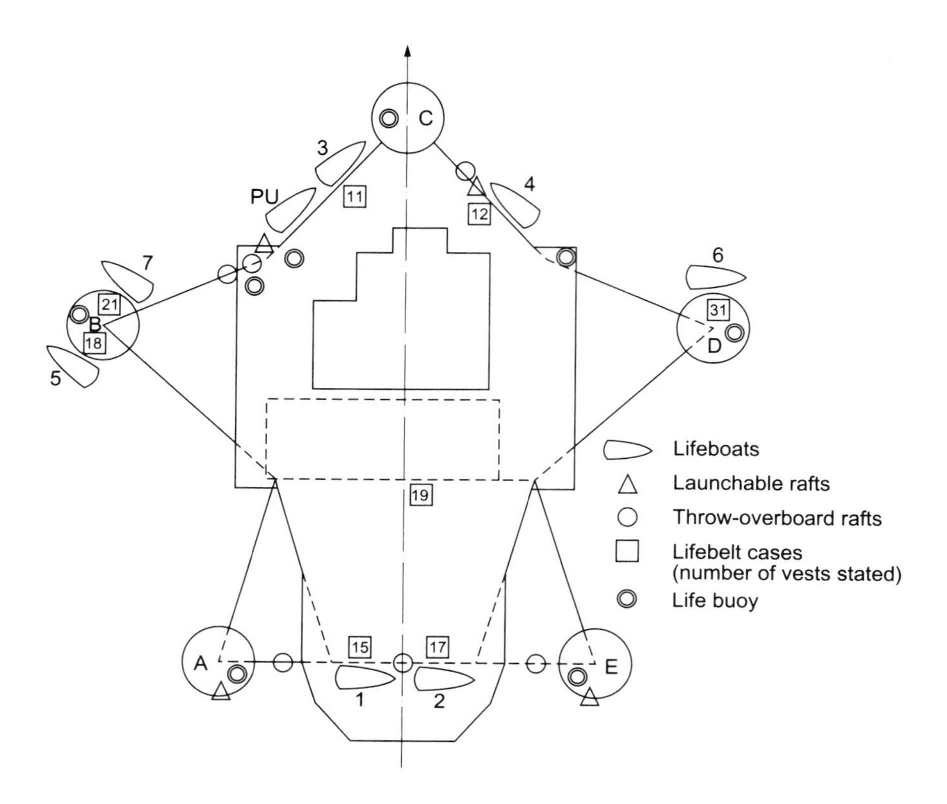

2.9 Diagrammatic plan view of *Alexander L Kielland*, showing location of rescue equipment.[9]

the position also at the time of the accident except that four of the launchable rafts were ashore for maintenance.

The accident

For nine months prior to the accident *Alexander L Kielland* had been anchored close to the production platform *Edda 2/7 C*. Anchors were attached to all columns except column C, which was located nearest the production platform. Contact between the two platforms was maintained by a moveable walkway. In bad weather this walkway was hoisted on board the *Alexander L Kielland*, which was then winched away from the *Edda 2/7 C* by slacking the anchor wires on columns B and D and tightening those on A and E.

On Thursday, 27 March 1980 the weather was indeed poor. The wind speed was 36–45 mph and the wave height 20–26 ft. Accordingly it was decided to move away from *Edda 2/7 C* by means of the winches. This was accomplished without incident by 5.50 pm. A few minutes before 6.30 those on board felt a sudden impact followed by trembling. At first this was thought to be due to a wave, but shortly afterwards there was a further impact and the platform shook and started to heel. The heeling continued until it reached an angle of 35°, at which point the radio operator sent out an emergency message 'Mayday, Mayday, *Kielland* is capsizing'. For a time the list stabilised, although the platform continued to take water through ventilators and other openings in the deck. One anchor wire, on B column, remained intact, but strained like a violin string. Eventually, at 6.53, a little more than 20 minutes after the collapse started, the wire snapped and the platform overturned and floated upside down in the water.

Evacuation and rescue

Immediately before the accident most of the men were either in the mess-hall or in the cinemas, and very few in the bedrooms, which were located aft. The heeling, which was due to the failure of the brace D6 and disintegration of column D, was a sudden collapse, which was arrested when part of the platform became submerged. As a result, most of those on board were thrown to one side of the room in which they had been sitting. From later medical evidence however, it would seem that only a few were seriously injured at this time. Most were able to make their way on deck, although with some difficulty.

Twenty-six men escaped aft, where lifeboats No 1 and 2 were located below the helideck. Because of the list, No 2 boat was underwater but No 1 was clear. Its engine was started and when no more people appeared, the boat was launched. The lifeboats were provided with a release wire, and after it was afloat, pulling this wire was supposed to open the hooks that attached the boat to the davits. This failed to operate, and while one man was pulling frantically at the release wire, another attacked the forward hook with an axe. Then suddenly the forward hook released itself, but by this time they had been thrown back on to the platform, with some damage to the superstructure. The back of the wheelhouse was stoved in, and whilst they were stranded one man was able to reach through the hole and release the other hook. The boat then floated clear of the wreck. It was seaworthy and the engine was operating, but radio communication was difficult. Eventually, at about 1.20 the following morning, they were spotted by two of the supply ships. However, since the lifeboat was in good shape it was decided not to transfer the men to these ships and they were rescued by helicopter, the last man being winched up at 5.00 in the morning. The weather continued to be bad, visibility was poor and throughout this protracted operation men were seasick and very cold.

Most of those who managed to escape from the mess room, cinemas and quarters made for the highest point, which was column B. Lifeboat No 5 was located here, but only 14 men got on board; others thought it would be crushed against the crossbrace as it was lowered. In the event it never was lowered, but was flung into the sea bottom up as the platform capsized. Fortunately the hooks were released and the boat drifted away from the upturned platform. Some of the survivors from the platform swam over to the lifeboat and by superhuman efforts, helped by those inside, managed to right it. A further 19 people were then able to board the craft. It was not possible to start the engine and the radio did not work, but at 7.30 in the evening the boat was found by the supply ship *Normand Skipper*. Twelve men were taken on board the ship by an entry net before the operation became too hazardous, and the ship stood off. The other men were rescued by helicopter between 2.30 and 4.00 in the morning.

Of the remaining lifeboats, No 2, which had been already submerged, went down with the platform and No 6 was lost when D column collapsed. No 3 was located on the port side, and was lowered with seven or eight persons on board. The launch was successful but it was only possible to release the aft hook. The boat was pushed under the platform but was then lifted up by a big wave and landed on top of the winch on C column. Those on board managed to escape and, except for one man who was injured, were eventually rescued.

No 4 lifeboat was lowered but was crushed, and there were no survivors. No 7 boat was launched but it capsized. A number of those aboard were able to get through the side hatches, and of these, three were rescued.

Altogether 59 survivors were picked up by helicopter from the lifeboats. The other 30 were rescued from liferafts, or swam to the operating platform, or were rescued by the supply ships.

The rescue helicopters and ships had serious difficulties owing to poor visibility and to a minor extent, because of problems with the winches on helicopters. Nevertheless, they picked up all those who were on rafts or lifeboats, whilst the ships and *Edda 2/7 C* rescued men from the sea. The number of rescue units increased during the operation, such that eventually the Norwegian authorities requested that an RAF Nimrod aircraft should co-ordinate air movements, whilst a Dutch warship controlled shipping. This arrangement seems to have worked very well.

The cause of the failure

It has already been mentioned that the *Alexander L Kielland* was originally designed as a drilling platform. As such, it was equipped with hydrophones. These instruments pick up an acoustic signal from a source located in the well, and enable the platform to be positioned accurately. It so happened that at the time of the accident, the platform was being prepared for drilling operations, so the hydrophones were by no means superfluous. These instruments were mounted on the lower horizontal braces and, in particular, there was one on brace D6 which, as shown in Fig. 2.8, joined column D with a node in the brace between column C and E. The location of this hydrophone is illustrated in greater detail in Fig. 2.10(a). The hydrophone itself was mounted inside a cylindrical fitting which was attached to the brace (Fig. 2.10(b)). The cylinder was made by rolling plate to the required diameter and joining the edges with a butt weld. A matching hole was cut out of the brace and the cylinder fixed in position by one internal and one external fillet weld, each of 6 mm throat thickness. The hydrophone and its mounting were classified by the designer as an instrument, and therefore no stress analysis was made of this detail.

The collapse started as the result of a fatigue failure of brace D6, and this in turn was initiated at the toe of the fillet welds. The crack started at the 12 and 6 o'clock positions, looking at the hydrophone fitting with the brace horizontal. It then propagated around the circumference of the brace until the remaining ligament was too small to sustain the applied stress, when it suffered a sudden catastrophic failure. It has been estimated that the parting of this ligament took place in less than one-hundredth of a second. As a result, the remaining braces connecting column D were subject to dynamic loading which had the effect of amplifying the loads to which they would have been subject under static conditions. These braces were not designed to operate safely in the absence of brace D6, so they failed, also in a very short time. Inevitably column D then collapsed. The failures were partly shear failures, partly flat fractures, and in most cases occurred in two places, close to the nodes or to the junction with the column. There is no suggestion that inferior material greatly contributed to these later failures; on the contrary, later testing of the material showed that it met specification requirements except for a few trivial deviations and was generally of good notch-ductility at the temperature then prevailing (between 4 and 6 °C).

The appearance of the fracture in the brace D6 is shown in Fig. 2.11. There were two independent initiation sites, No I from the outside fillet weld and No II from the inside fillet. Such points of initiation of fatigue cracking are typical of a fillet-welded attachment exposed to an alternating

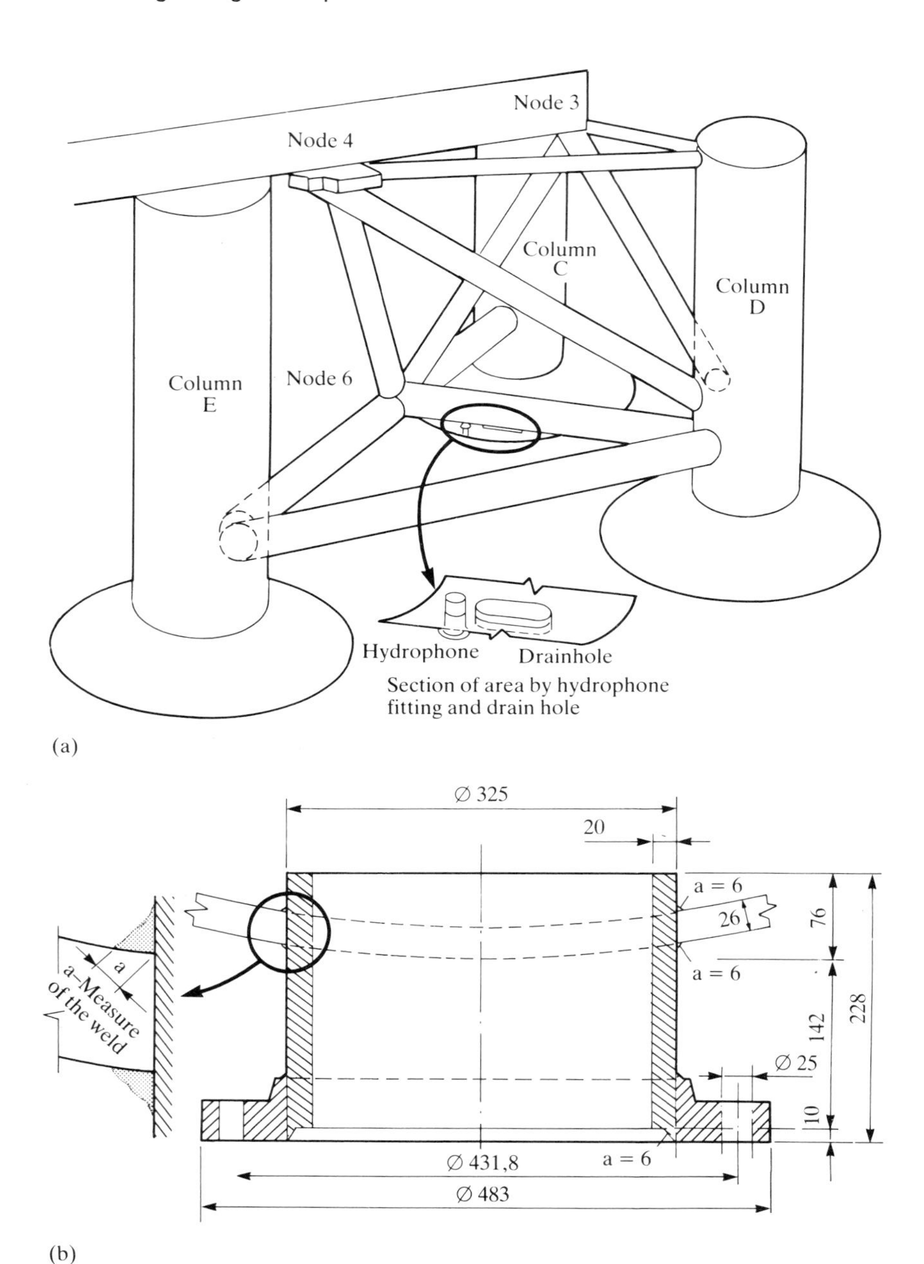

2.10 The hydrophone fitting on brace D-6 of the *Alexander L Kielland:* (a) location of fitting; (b) sectional elevation with nominal dimensions in mm. The dimension *a* is the throat thickness of the attachment weld, specified to be 6 mm.

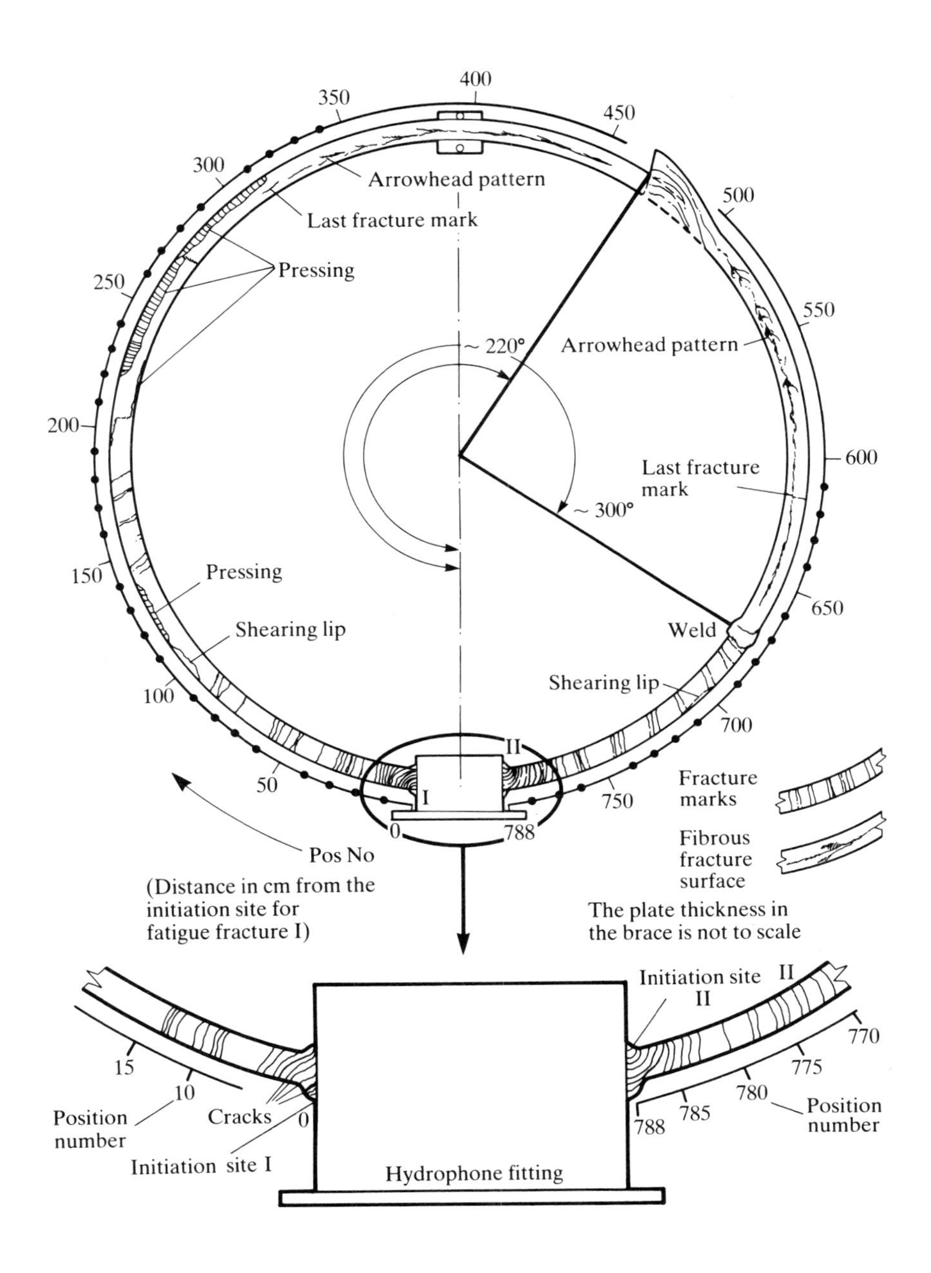

2.11 Surface appearance of fracture in brace D-6 of the *Alexander L Kielland*.[9]

tensile stress. Fillet welds reduce the fatigue strength of steels in two ways. Firstly, the profile of the weld causes a stress concentration. Secondly, a crack or slag-filled intrusion, commonly up to 0.5 mm in depth, is usually to be found near the toe of the fillet. This defect is responsible for the major part of the reduction of fatigue life, and it is current practice to grind the toe of fillet welds in critical areas to remove such defects.

The general appearance of the first part of the crack, as shown in the sketch, is characteristic of a fatigue failure. There is little or no reduction in thickness, and the surface is marked periodically with beachmarks. These represent either a slight change in direction, or a change in the crack growth rate. Examination under the microscope shows additional features, in particular there are fine striations which are generally assumed to indicate individual load cycles.

After the first 300 mm or so on either side of the hydrophone fitting the fracture appearance changes, indicating that it was growing in leaps, giving a coarse and fibrous fracture surface. Finally the last third of the circumference has a woody appearance with chevron (herring-bone) marks which are typical of fast, unstable crack propagation. These various modes of failure are discussed more fully in Chapter 3. There were also cracks all around the circumference of the cylinder that formed the hydrophone mounting. Some of these cracks had paint on the surface, from which it may be deduced that they occurred during the welding operation. Most of the cracks, including those that were contaminated by paint, had a rough fibrous surface and lay about 1 mm below the plate surface. Such fractures are typical of the defect known as 'lamellar tearing' and are caused by a combination of shrinkage strain during cooling of the weld with low through-thickness ductility in the steel. This ductility was measured for the hydrophone fitting as between 1 and 7%, which indicates a very low resistance to lamellar tearing. The implication of these observations and tests is that for much, and probably most, of its life the fitting was only connected to the brace over part of the weld area. The result would have been to increase the level of strain around the fitting and to give a strain concentration effect in the regions where the fatigue crack initiated.

Various calculations have been made subsequently to determine the probable life expectancy of the brace D6. Assuming a sound weld between the hydrophone fitting and the brace, and using the data then available concerning wave frequency, it was estimated that the lifetime would be between 24 and 150 months. If the fitting had been completely separated, this estimate was reduced to between 10 and 54 months. The actual life was just over 40 months, so these estimates are reasonably close to the mark.

It would be reasonable to conclude that the most important factor leading to the premature failure of brace D6 was the use of low quality steel for the hydrophone fitting. A preliminary examination of one of the other fittings showed no surface cracking in the exterior weld, and there was no evidence of incipient failure at other joints. Thus, although the design was suspect, defective material would seem to have been the reason why a potential for fatigue cracking became a catastrophic failure.

The historical perspective

When a mechanical failure gives rise to a disaster of such magnitude as that which befell the *Alexander L Kielland*, then the adequacy of the design and control system is inevitably brought into question. In this case there were two organisations mainly responsible: the original designers, Institut Français du Petrole and Neptune, and the classification society responsible for reviewing the design and carrying out inspection during the construction phase, det Norske Veritas. It will be recalled that the design of Pentagone rigs 82–91, including No 89, *Alexander L Kielland*, was essentially the same, and based on P82, which was designed in 1970–71. P89 was delivered in July 1976.

The 1970s were a period of considerable development in the understanding of how far, in quantitative terms, the presence of fillet-welded details affected the fatigue strength of steel. The fact that welding reduced the fatigue life had been known for a number of years and Gurney published a book on the subject in 1968.[10] In the early 1970s data on fatigue strength was gathered by the British Welding Institute and this resulted in proposals for design rules which appeared in 1976.[11] These proposals were incorporated in Part 10 of the British Standard for steel bridges BS 5400 in 1980. However, none of the classification societies had incorporated any provision for design against fatigue in their rules by 1976. There was therefore no reason to expect either Lloyd's Register (which was concerned with the original design) or det Norske Veritas to have carried out a fatigue analysis. At the time it was generally assumed that the fatigue limit of carbon steel was about half the tensile strength, so any safety factor greater than two would take care of the problem.

Likewise there was no requirement for control of the through-thickness ductility of steel plate. Lamellar tearing had been encountered, but mostly in heavy plate fabrication and its potentially dire consequences in the welding of a detail such as the hydrophone mounting was certainly not understood at the time. It was only after the accident that it became normal practice to specify a minimum through-thickness ductility for steel used in critical joints; a typical minimum value in current practice is 20%.

The other design fault was that there was no redundancy in the structure; that is to say, if one major load-carrying member failed, others were overloaded and a collapse or partial collapse would be inevitable. At the time that the Pentagone rigs were developed it was already standard practice in some areas of engineering to protect against such an outcome; for example, this was done in aircraft, where it was known as 'fail-safe' design. In some cases there is a built-in redundancy; for example, the hull plating of a ship provides an additional measure of support. Not surprisingly the commission of inquiry recommended that semi-submersible rigs be designed to withstand the failure of a single member.

Escape craft

The difficulty and danger of launching lifeboats in heavy seas was undoubtedly one of the important factors leading to the heavy loss of life in the *Alexander L Kielland* accident. A further contribution was the difficulty in releasing the hooks which, in effect, tied the lifeboats to the wreck. It is vital that as soon as the boat is afloat the forward and aft hooks should be released simultaneously. The motion of the boat made this impossible in the case of the *Alexander L Kielland* because the hook could only be released when there was no tension, and when the boat rocked one hook was loose whilst the other was on load. The Commission of Inquiry could find no solution to this problem, because some time previously the release wire of a survival capsule had been operated prematurely and the capsule fell, killing three men. Subsequent to this tragedy, it was decided that hooks should not be releasable when under load, and the Commission of Inquiry could not recommend any relaxation of this rule.

In other countries such a relaxation has been adopted and it is possible for those inside the boat to release the hooks simultaneously. A hydraulic interlock ensures that this can only be done when the boat is actually afloat. The Norwegian authorities have, however, opted for a much more radical solution: the free-fall lifeboat. The boat dives into the sea from a height of up to 30 m; there are no davits and no hooks, and once released the boat goes in the right direction, away from the wreck. Occupants of free-fall lifeboats lie prone on specially contoured beds and both body and head are restrained by straps. Where free-fall lifeboats are installed, personnel are trained in their use. This training includes at least one dive in a free-fall boat. To date the system works well during training, but it has yet to be tested in an emergency.

There were 212 men on board the *Alexander L Kielland* when the supporting column D collapsed. Of this complement 89 were rescued, and 123 lost their lives.

The Piper Alpha disaster

The Piper field which produces both oil and gas was, at the time of the accident, operated by the Occidental Group and is located just over 100 miles northeast of Aberdeen, Scotland, in the North Sea. Figure 2.12 shows how the pipelines connected the three fixed units in the field, the gas compression platform MCP-01, and the two terminals, the oil terminal at Flotta in the Orkney Isles and the gas terminal at St Fergus in Scotland. Initially only oil was exported, gas which was surplus to platform requirements being flared, but in 1978 following governmental requirements for conservation, gas was exported to MCP-01 where it was mixed with gas from the Frigg gas field.

Fig. 2.13 is a photograph of the platform taken from the north-west, showing the accommodation with the helideck on top and the radio room on the far side of the helideck. The drilling derrick is at the rear. This rig provided means for drilling wells to the producing reservoir, together with process equipment to separate oil, water and gas and to separate gas

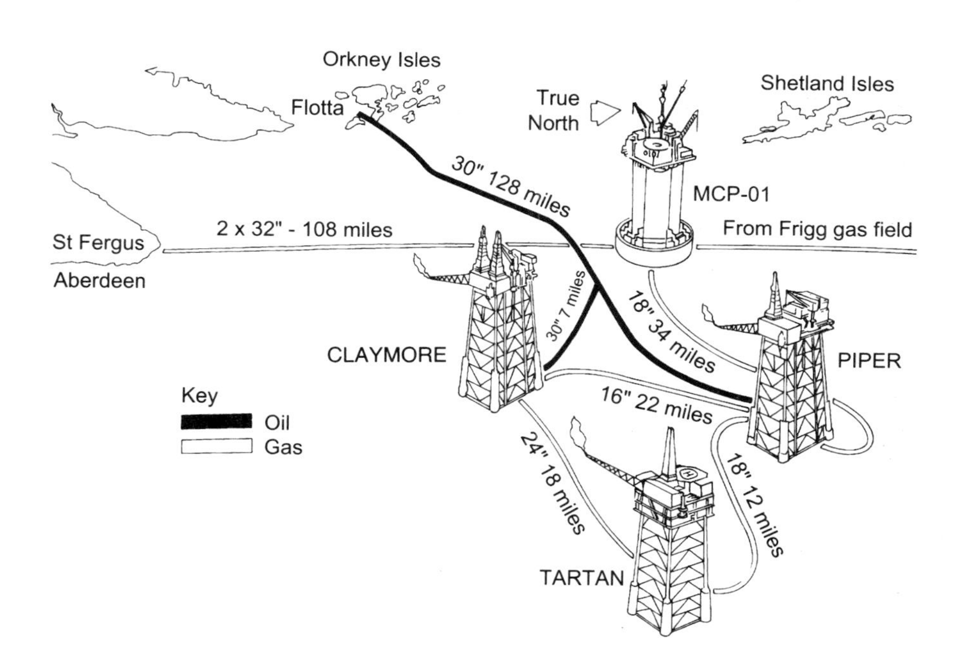

2.12 Pipeline connections in the Piper field in 1988.[12]

2.13 The Piper Alpha platform view from the north-west.[12]

(mainly methane) from condensate (mainly propane). Further processing of the gas and condensate was carried out on shore at the St Fergus terminal.

The jacket was a steel structure standing in water of depth 474 ft. There were decks at 20, 45, 68, 84, 107, 121 ft, then four levels of accommodation and the helideck at 174 ft. The production deck, which housed most of the processing equipment, was located at the 84 ft level, as outlined in Fig. 2.14. The area was divided into modules, each about 150 ft long, 50 ft wide and 24 ft high. On the left in Fig. 2.14 is A module, which contained the wellheads or 'Christmas trees' (so called because of their shape), of which there were 3 rows of 12.

The B module housed the two production separation vessels and one smaller test separator, together with the pumps for the main oil line. Module C contained the gas compressors with their associated scrubber vessels and coolers. There were three centrifugal and two reciprocating compressors. This is where the trouble started. D module was primarily for power generation. The gas conservation module, a later addition, was located as shown on the next level up, and below, on the 68 ft level, were the terminations of the gas lines from the Tartan and Claymore platforms and the gas line to the MCP-01 compression unit. This level contained the

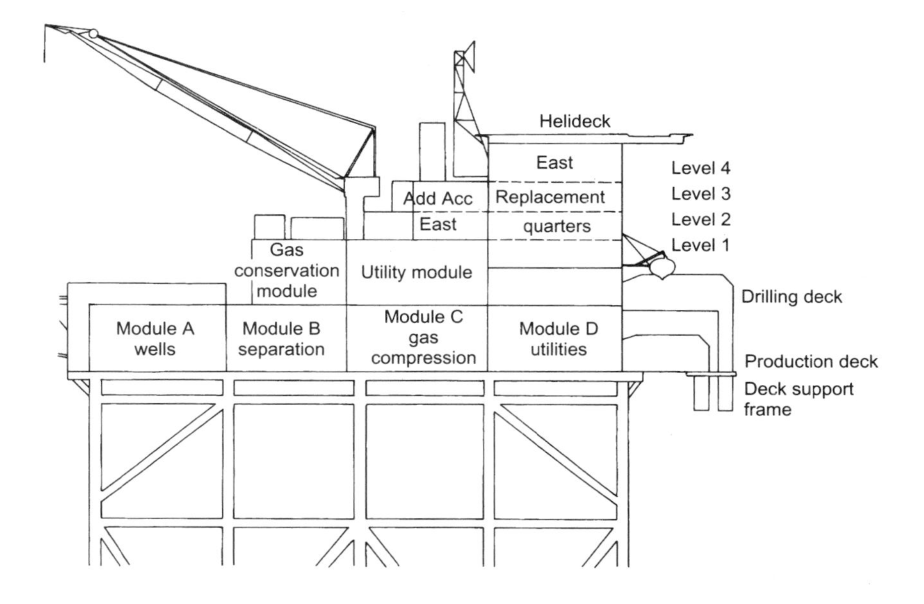

2.14 Layout of the topsides of the Piper Alpha platform.[12]

condensate injection pumps and the drum in which gas and condensate were separated. The diving area was also situated at the 68 ft level.

The control room was on a mezzanine floor, above the D module. In current practice, virtually all the operational functions of a process unit can be carried out in the control room, either by human beings or by computer. The Piper Alpha control room was an earlier type, where conditions in the plant were monitored and from which instructions to operators in the relevant module were issued. For example, if a high level alarm showed up for a particular vessel, the control room operator telephoned the appropriate module operator to cut the flow to that vessel, which he did from a local control board.

The process unit

Figure 2.15 is a simplified process flow diagram for the unit as it was being operated at the time of the accident. The first step is primary separation, by gravity, of water, oil and gas. The water passed through a hydrocyclone to remove oil and was then discharged into the sea. Oil (which at that stage usually contained about 2% water) was pumped directly to Flotta. The gas, which consisted of a mixture of gaseous hydrocarbons, was compressed to 675 psi in three parallel centrifugal compressors, and

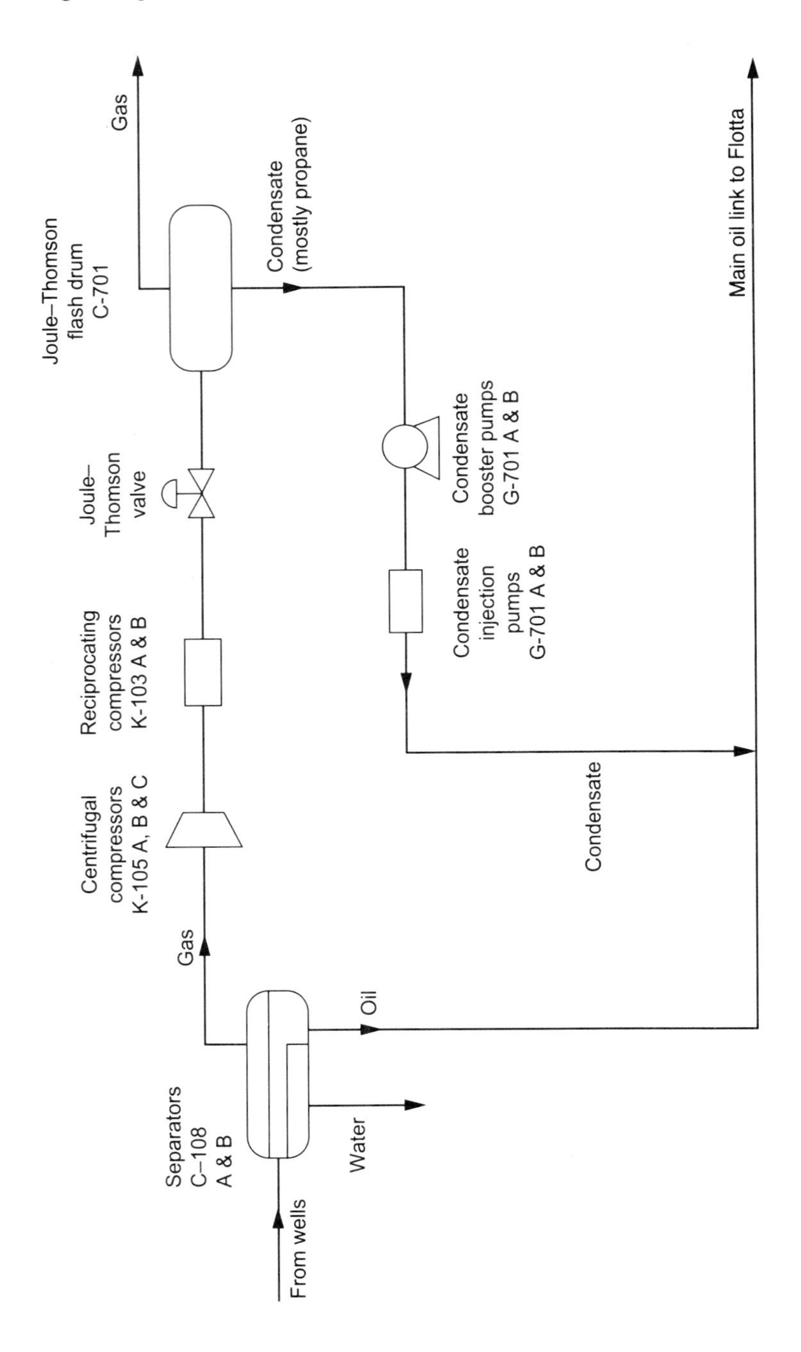

2.15 Simplified process flow diagram for Piper Alpha.

then to 1735 psi by two parallel reciprocating compressors. In the operating mode in use at the time of the accident, the gas was then chilled by passing it through a Joule–Thomson expansion valve, where the pressure was reduced. The resulting fall in temperature caused higher hydrocarbons such as propane to condense, whilst lower hydrocarbons such as methane remained gaseous. These two phases were separated in a flash drum. The condensate was then compressed and injected into the main oil line, to be separated again at the terminal.

In the other operating mode gas from the reciprocating compressors passed to the gas conservation module, where it first went through molecular sieve driers then to a turbo-expander where the pressure dropped to about 635 psi. The condensate so formed went to a demethaniser tower where methane was stripped out, and thence to the Joule–Thomson flash drum. Gas rejoined the original system downstream of the same flash driers.

Power

The main electrical supply came from two 24 000 kW generators driven by gas turbines. Normally the turbines were fired by fuel gas but they could alternatively operate with diesel fuel. In the event of a fall in gas pressure there was an automatic switch to diesel. However, this changeover was said to be less than 100% reliable. There was a completely separate power supply for drilling.

If the main supply failed, than a diesel-fired emergency generator came into operation. As back-up, and to supply power to essential items whilst the emergency generator was running up to speed, there was a battery-powered supply.

6 July 1988

A contractor was in the course of calibrating some of the safety valves in the Piper Alpha process equipment. On this date the last of the series, PSV504 on the condensate injection pump G701A, which had already been shut down and electrically isolated in preparation for other maintenance work, was to be checked. As noted earlier, the condensate injection pumps were on the 68 ft level, but the relief valve was located above, in the corner of C module. The valve was removed in the early afternoon and a blind flange fitted in its place. It was duly tested and found satisfactory, but could not be replaced because there was no crane available. It was therefore agreed with the day-shift operators that the work should be completed on the following day.

At about 9.45 the same evening, after the night-shift had taken over, the remaining condensate injection pump, G701 B, tripped out. This meant that the flow of condensate out of the Joule–Thomson flash drum was arrested. The control room operator informed the operator in C module and at the same time the lead operator, who was in the control room, went down to try to restart the pump. Shortly after this, a high level alarm showed up for the surge drum and the reciprocating compressors were put on recycle. It would then only be about half an hour before the gas supply to the main generators failed and the changeover to diesel fuel initiated. The operators were therefore under some pressure.

At this point the lead operator came back to the control room to report that the B compressor would not restart. The operators were, it would seem, unaware of the fact that the safety valve had been removed from the A compressor, and they proceeded to cancel the 'permission to work' permit so that it could legitimately be reconnected to the electrical supply. The lead operator (who died in the disaster) then went back to the compressors.

Shortly after this, two of the centrifugal compressors tripped, and more or less at the same time a low level gas alarm in C module was activated. Then things happened very quickly. A further set of gas alarms sounded in rapid succession, three low level and one high level, all in C module. The control room operator was trying to talk to the operator in C module above the noise of these alarms when the first explosion occurred. The time was 10.00 in the evening.

This was clearly a vapour cloud explosion, but the source of the vapour cannot be established with complete certainty. Nevertheless, the evidence assembled by Lord Cullen and the experts who conducted the public inquiry is very convincing. The lead operator, having signed off the permit to work on the G701 A pump returned to the 68 ft level, with the intention of arranging for the electric power to be reinstated. The first step in restarting the pump would be to open the valve on the suction (upstream) side and to admit condensate. From other evidence it is supposed that this action was taken, and that the valve was opened for about 30 seconds. This would have had the effect of admitting condensate to the pump and to the relief valve line at a pressure of 670 pounds force per square inch absolute.

According to local practice, there were three levels to which the bolt fastening the blind flange to the relief valve piping could have been tightened: finger tight, hand tight (using a spanner) or flogged (using a flogging spanner). Tests subsequently showed that the second two levels of tightness would have made a leak-tight joint, but that a finger tight joint would leak. Condensate escaping in this way would partially vaporise, and would have provided a sufficiently large vapour cloud to account for the explosion. Other sources of vapour were considered to have been very much less probable.

Immediate effect of the explosion

Between C module and its neighbours, B and D, there were fire walls. These consisted of sheet steel bolted to a steel framework, and insulated with mineral wool or other heat insulation material. The force of the explosion was sufficient to blow these walls out, and cause fatal damage in the adjacent modules. In D module the main and emergency electrical systems and the control room were extensively damaged and air lines torn out. Thus all means of controlling the platform as a whole were lost in a fraction of a second. However, the system operated like the vacuum brakes on a vehicle; loss of air and power caused units to shut down and emergency valves to close. The flare continued to operate and would eventually have depressurised all or most of the units.

However, in B module the explosion had devastating effects. Disintegration of the fire wall provided flying debris which damaged pipework and caused a leak. This is thought to have occurred in the 4 in line carrying condensate just upstream of the point where it joined the main oil line (see Fig. 2.15). This leak developed rapidly into a full-bore fracture, the condensate discharged with great force, causing a fireball, and crude oil started to pour out of the main oil line, forming a large spreading fire. The cause of the full-bore fracture is not known, but experience shows that where there is a leak in a gas line due to a sudden local crack, then an internal explosion can occur in the pipe, blowing it wide open. However this may be, the scene was set for a disastrous escalation.

A major source of fuel for the spreading fire was the separators, which contained about 50 tons of crude oil. Nevertheless experts calculated that this would not be sufficient to account for the known degree of spread, and it seems likely that there was an additional source. There was an emergency shut-down valve in the main oil line which should have operated when power was lost; however, such valves do not always close fully, so a leak was possible. Pressure was maintained in the main oil line because oil production continued at both the Claymore and Tartan platforms. It was not until about an hour after the initial explosion that a shore-based manager instructed Claymore and Tartan to shut down and made arrangements for the terminal at Flotta to depressurise the line.

The fire had meantime spread down to the 68 ft level, where the three gas risers terminated. There were two import risers, one from Tartan and one from Claymore, and an export line to the compressor platform MCP-01. These were large diameter high tensile steel lines; for example, the Tartan riser was 18 in in diameter with a 1 in thick wall, and the normal gas pressure was about 1700 psi. The risers and their associated pipelines contained large quantities of gas, such that to depressurise them by flaring off the contents took several days. In all cases the valve at the 68 ft level had closed after the initial explosion so those parts of the risers exposed to the fire were at full pressure.

At 10.20 pm the Tartan riser burst, causing a major explosion and engulfing the platform in a sudden and intense fire. The rupture was caused by the direct exposure of the pipe to fire, such that the strength of the steel was reduced. Under normal circumstances the riser would have been protected by a fire-water deluge, but the fire-water system had been disabled by the earlier explosion.

Half an hour later there was a third violent explosion, the vibration of which was felt a mile away. This explosion destroyed one of the rescue craft operating near the platform and killed all but one of the crew. It was almost certainly due to the rupture of the MCP-01 line.

At this stage the platform started to collapse. The jib and cap of the crane on the west side fell into the sea. Shortly after this there was a major structural failure in the centre of the platform. The drilling derrick fell across the pipe deck, which collapsed. At about this time, approximately an hour after the failure of the Tartan riser, there was a further major explosion which was probably due to rupture of the Claymore gas riser. The drill store, where a number of men had taken shelter, started to fall, forcing the men out and engulfing some in smoke and flames. The supports for the accommodation modules began to fail and one by one they fell into the sea, that on the north end overturning at 00.45 am. By this time the fire was mainly from some of the wells and from oil floating on the sea, the risers having disintegrated down to near sea level. Figure 2.16 shows the remains of the platform on the morning of 7 July, with flames still coming from one of the risers (presumably the MCP-01 line, which was 128 miles long – see Fig. 2.12). The black smoke is from the burning wells, seven of which were still flowing.

Evacuation and rescue

The emergency procedures for the rig required that the crew should muster in the accommodation area, where an emergency evacuation squad would direct parties to their respective lifeboats or to the helicopter. The whole operation was to be directed by the Offshore Installation Manager. In the event these arrangements were completely set at nought by the

2.16 The remains of the Piper Alpha platform on the morning of 7 July. One of the risers is still burning (on the right of the picture). The black smoke is from the burning wells.[12]

failure of the public address system, such that no general directions could be given, and the rapid spread of the fire, which made access to the lifeboats impossible. It was clear to some of those on board at an early stage that smoke would make a helicopter landing difficult.

For those on the night shift there was no information at all, and they had to make their own decisions. The divers, after recovering from the shock of the initial explosion, cast about for escape routes and found none, nor was there any way to get to their lifeboat. One of the team was working undersea; he was recovered and spent a short time in the decompression chamber. They then made their way on the 68 ft level to the north-west corner of the rig, which was fairly clear of smoke. In this location there were knotted ropes hanging over the edge of the platform, and the men used these to climb down to the 20 ft level. Here they found rings attached to one of the platform legs and were able to climb down and step into one of the rescue boats. Others who were working on the 84 ft level went to the same point, including the two control room operators and twelve other men. Two of these fell off the ropes, and others were forced to jump into the sea from one level or the other.

Most of the drillers were able to go up to the accommodation level in accordance with the emergency procedure. There were already about 100 men in this area, and eventually most of them made their way to the galley. Here the conditions were at first tolerable and there was hope either of a helicopter rescue or that the semi-submersible *Tharos*, which was nearby, would be able to get them off by a walkway. However, after the explosion of the Tartan riser at 10.20 pm many decided that the only way to be rescued was to get off the platform, and found their way by various routes out of the accommodation. Seven went up on to the helideck, and when the second major explosion occurred four of these jumped into the sea, 175 ft below. They all survived, as did one other who jumped off the helideck later. Others went downwards and got off the platform into the sea by one means or another. Altogether there were 71 survivors. Sixty-three per cent of the night staff survived but only 13% of those off duty. When the accommodation units were recovered from the sea-bed, 81 bodies were found inside. Altogether 165 of the 226 men on board lost their lives.

At the time of the disaster there were four vessels close to Piper Alpha. The standby was a converted trawler, the *Silver Pit*, and it lay 250 m north-west of the platform. This vessel carried a fast rescue craft, a diesel-driven water-jet boat capable of 30 knots, and carrying 3 crew and up to 12 survivors. At 556 m was the *Tharos*. This was a semi-submersible which was equipped for fire-fighting, had a hospital with 22 beds, a fast rescue craft and a helicopter. She also had means of launching a walkway on to a platform to evacuate crew. The *Maersk Cutter* was a supply vessel located about a mile from Piper Alpha. This had a fire monitor capable of discharging 10 000 tons of water per hour. The *Lowland Cavalier* lay 25 m off the south-west corner of the platform, engaged in trenching operations. Other vessels in the vicinity were the *Sandhaven*, a converted supply ship carrying a petrol-driven fast rescue craft, the *Loch Shuna* and the *Loch Carron*, also supply ships.

The *Tharos* started to move towards Piper Alpha immediately after the first explosion. Her helicopter was airborne at 10.11 pm but the pilot reported that the helideck was obscured by smoke. This helicopter was not equipped with a winch and it took no further part in the rescue operation. Preparations were then made to provide a cascade of water. This cascade came into action after some delay and provided a certain amount of protection against heat for the rescue boats and others. There was also an attempt to deploy the gangway. However, the landing position on the platform was shrouded in flame and smoke and the attempt was abandoned. Eventually when the MCP-01 riser exploded, *Tharos* was partly enveloped by the fireball and was drawn back by 100 m. Meantime the *Maersk Cutter* had been directing water jets at the drill floor and continued to do so until shortly after midnight.

Fast rescue craft had been launched from a number of vessels and these were mainly responsible for picking up survivors. The first was from the *Silver Pit*, and it was this boat that picked up the divers, amongst others. It carried on working until after midnight, when the hull and engine were damaged by a further explosion. Its occupants were rescued by the *Maersk Cutter*. The fast rescue boat from the *Sandhaven* had picked up four men from the south-west corner of the platform when the MCP-01 riser ruptured. The explosion and fireball destroyed the boat. Two of the crew and all those that had been rescued were killed.

Of the 61 survivors, 37 reached the *Silver Pit*, 29 having been picked up by the rescue boat and the remainder by the vessel itself. Others were taken by rescue boats to the *Tharos* or were picked up directly by this vessel or the *Maersk Cutter*. No lifeboats were launched from Piper Alpha and there were no helicopter rescues.

Essential features of the Piper Alpha disaster

The series of explosions and fires on board Piper Alpha arose because an operator was attempting to start up a condensate compressor, and as a preliminary step, pressurised the pump and the relief valve line. He was aware that the pump had been electrically isolated preparatory to maintenance work but (it is virtually certain) did not know that the safety valve had been removed for calibration and had not been replaced. The safety valve was located at a higher platform level than the compressor and was not visible from the compressor itself. Pressurising the pump caused a leak of condensate leading to a vapour cloud explosion which destroyed all means of control on the platform and initiated a major fire, fuelled by crude oil from the separators and also possibly from the main oil line. As a result of the loss of power all emergency shutdown valves closed, including those on the import and export gas risers. These shutdown valves were located at the platform level which was seriously affected by the fire. The piping was weakened by heating in the fire, and three of the risers burst one after the other. Because these risers were connected to lengthy pipelines (in one case 128 miles long) and because the gas was under high pressure, there was an intense fire that persisted for a long time and eventually destroyed the platform.

The high loss of life was due eventually to two factors: first there was no means of escaping the potentially deadly effects of smoke, and secondly there was no means of access to lifeboats or other orderly means of escape. Only those who were bold enough to jump into the sea or lucky enough to be near an easy escape route survived.

The South Pass 60 Platform 'B' fire

In March 1989, less than a year after the Piper Alpha catastrophe, there was a similar type of accident in the Mexican Gulf off the coast of Louisiana. Platforms in the Gulf are normally much smaller than those in the North Sea. The unit in question had 11 risers, one carrying oil, one condensate and the remainder gas. On 19 March a contractor's crew were cutting an 18 inch gas riser in preparation for installing a pig trap. They had just penetrated the pipe wall when condensate started to spray out. The vapour was ignited by nearby machinery and a considerable fire ensued, the gas pressure in the riser being 1000 psi. The emergency shutdown system came into operation and, in particular, valves on all the other risers closed. As in the case of Piper Alpha these valves were above the source of the fire, and eventually six of the risers burst and the explosions and fires destroyed the platform, which finished up in much the same condition as Piper Alpha. Not only were the mechanics of this accident similar, but the action which initiated it resulted from the same human cause; a failure of communication, and in particular a failure to plan the piping modifications so that they would be safe and that all concerned were aware of potential hazards.

Seven men died as a result of this disaster. The US task group that was given the job of reviewing the Piper Alpha and the South Pass 60 incidents concluded that if the workforce at South Pass had been of similar size, then the casualties could have been as high as those on Piper Alpha.[13]

Protecting the gas risers

The problem about gas is its compressibility. The compressibility of liquids is small so that a pipeline carrying liquid hydrocarbons can be depressurised by drawing off a small fraction of its total content, and this can be accomplished in a short space of time. To depressurise a gas line operating at 3000 psi (not an uncommon pressure in offshore work) by flaring to atmosphere could require the burning of about 20 times the volume of the pipeline, which would take a long time. Likewise, a fire following a full-bore rupture persists for a period long enough to cause structural collapse of the surrounding steelwork.

One way of reducing the risk of failure due to an external fire is to insulate the riser. This was done in the case of one of the Ekofisk platforms situated in the Norwegian section of the North Sea, but corrosion occurred under the insulation, and eventually the pipe leaked and there was a fire. It was during the evacuation of this platform that the hooks of a lifeboat were accidentally released, killing three men. And this was the incident that caused the Norwegian authorities to specify the type of hook

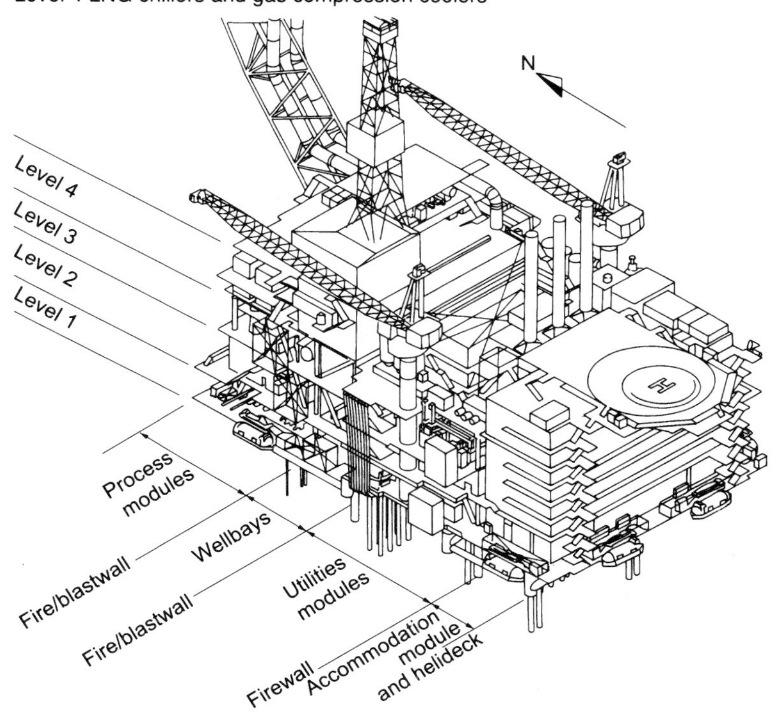

2.18 Diagram of the topsides of the Tiffany platform, incorporating safety features.[16]

The Comet I aircraft failures

These early jet aircraft first took off a long time ago, and the company that made them no longer exists. Nevertheless, there were important technical lessons to be learnt from their loss, lessons that are not always fully understood today, so the history of the disasters is still worthy of study. Fortunately a very lucid account of the accidents was provided by the court of inquiry which met under the chairmanship of Lord Cohen.[17]

Comets were built by the de Havilland Aircraft Company and the engines were supplied by a subsidiary, the de Havilland Engine Company. de Havilland was founded immediately after World War I by Captain Geoffrey de Havilland. It manufactured civil aircraft, mostly airliners, which in the early days were biplanes powered by a single engine. They were numbered successively, the 1920 version DH 16, 1924 DH 34 and so on. Comet was DH 106. During World War II the entire output of the company consisted of military planes, but in 1945 it reverted to the civil field. The engine company had experience with gas turbines, so it was decided to take a bold step and design a jet airliner. At that time communications within the British Empire offered a lucrative trade; flights were long, and a jet aircraft could halve the flight time.

Design work on the prototype started in September 1946. At the beginning of the following year the British Overseas Aircraft Corporation (BOAC) and the Government Ministry concerned signed a contract for the purchase of Comet aircraft, and this enabled de Havilland to go ahead with production. Two prototypes were delivered in 1951 and BOAC started providing flights, having been issued with temporary certificates of airworthiness. By 1952 a full airworthiness certificate was obtained and a passenger service was started, the first flight being to Johannesburg in South Africa.

Design

Turbo-jet engines consume fuel at a much lower rate at high altitudes than they do lower down, and for this reason it was necessary that the Comet should be designed to fly at 35 000 ft or more. For the comfort of passengers and crew it was, therefore, necessary to pressurise the cabin. An excess pressure P (difference between internal and external pressure) of $8\frac{1}{4}$ psi was required, and this was 50% higher than that of any other aircraft operating at the time. Initially the design was based on static loading. The International Civil Aviation Organisation, and also the British airworthiness authority (the Air Registration Board), required that the maximum working stress should be less than half the ultimate strength of the material, and that the cabin should show no permanent deformation at an internal pressure excess of $1\frac{1}{3}P$. de Havilland went further, and used a design stress of $\frac{2}{5}$ ultimate strength and a test pressure of $2P$. This, they considered, would take care of any fatigue problems. They were also concerned to have a good margin of safety against the failure of individual windows, doors and hatches; with good reason because sudden depressurisation can cause death or injury to passengers and crew. Two test sections of cabin were built. The first part extended from the nose to the front spar of the wing, where it was sealed to a steel bulkhead. The second part extended 24 ft from just in front of the wing to a small distance aft of the wing. Both included typical windows, hatches and doors. These test sections were subject to 30 applications of pressure of between P and $2P$, and 200 pressurisations of just over P. Such tests were not intended to check the fatigue resistance, but were simply repeated static tests. Fatigue tests were, however, carried out on the wings, because fatigue cracking had been found in the wings of certain transport aircraft some time previously.

In 1952 it became evident from experience with military aircraft that fatigue failure could also occur as the result of repeated pressurisation of the cabin. The Air Registration Board therefore proposed that in addition to the static test, there should be a repeated loading test of 15 000 cycles at a pressure of $1\frac{1}{4}P$. In the middle of 1953 de Havilland decided that it was necessary to carry out such tests on the Comet cabin. They used the test section that ran from the nose to the wing, and subjected this to repeated pressurisations at the working pressure P, in addition to the static pressure tests that were done previously. The repeated loading test was discontinued when a crack, which originated at a defect in the skin, appeared near the corner of a window. By this time the pressure had been applied 18 000 times, and all concerned felt assured of the safety of the cabin.

The disaster

There had been a series of take-off incidents with the Comet in the early days, which culminated in a crash and fire at Karachi airport. These were originally put down to pilot error but further investigation showed that they were due to the wing section, and the problem was overcome by making some modifications. Then in May 1953 a Comet in flight from Calcutta to Karachi broke up in the air during a violent storm. On examining the wreckage it was found that the tail portion had broken away. It was concluded that in trying to counter the severe turbulence associated with the storm, the pilot had imposed loads that the airframe could not withstand. The earlier Comets were fitted with power controls so that it was possible to apply considerable force without being aware of the fact. Subsequently a synthetic resistance was applied to such controls which enabled the pilot to judge how much force he was applying to control surfaces.

Following these modifications all went well until January 1954, when Comet G-ALYP took off from Rome airport on a flight to London. At about 9.50 that morning another BOAC aircraft received a message from the Comet: 'George How Jig from George Yoke Peter did you get my' – at which point the message broke off abruptly. At this time the Comet would have been at about 27 000 ft, and still climbing. Ten minutes later farmers on the island of Elba saw aircraft wreckage, some of which was on fire, fall into the sea. The harbourmaster at Portoferraio was informed and he very promptly assembled a search and rescue team and set off for the area where the wreckage had been seen. Fifteen bodies, some mailbags and some floating wreckage was recovered. At the time of the accident, the aircraft was carrying 29 passengers and 6 crew, all of whom were killed.

It was evidently not a repeat of the Calcutta incident because the weather on this occasion was clear with little turbulence. The water was 400–600 ft in depth so salvage was possible, and the British Navy was given the job of recovering as much material as possible. Vessels were fitted with grabs and heavy lifting gear. They used a television camera underwater (the first time television had been employed in such an operation) to locate the wreckage. After about three weeks' search the remains were located and a fair proportion was recovered.

In the meantime all Comet flights had been suspended, and a committee of investigation had been set up under the chairmanship of Mr Abell of BOAC. By that time the fatigue test on the wings had generated a few cracks so that it was decided to strengthen the affected parts. The possibility of fatigue cracking of the cabin was discounted in view of the tests at de Havilland, and the committee took the view that fire was the most likely cause of the accident. Following this conclusion there was no reason why services should not be resumed, and so they restarted on 23 March 1954.

On 8 April of the same year a Comet took off from Rome airport on a flight to Cairo. Just over half an hour later when the aircraft would have been close to or at its cruising height, radio contact was lost. The following day bodies and wreckage were recovered from the sea off the Italian coast near Naples. It was obvious at once that the design of the aircraft was seriously flawed, and it was grounded again, this time permanently.

The similarity of the two accidents, in particular the fact that they had apparently broken up just before reaching cruising altitude, suggested that the cabin might have exploded because of the presence (against previous evidence) of fatigue cracks. Accordingly a water tank large enough to contain a complete aircraft was constructed and used to carry out fatigue tests on one of the Comets from the BOAC fleet. These tests were conducted with the cabin completely submerged so that there was no additional stress imposed by the weight of the water. The cycle consisted of pressurisation up to the normal excess of $8\frac{1}{4}$ psi, combined with loading of the wings to simulate flight conditions. Every 1000 cycles a pressure of 11 psi was applied. The aircraft used in this test had made 1230 pressurised flights, and its cabin failed after 1830 pressurisations in the tank, making a total of 3060. The first service failure, at Elba, occurred after 1290 flights, and the second after 900 flights. At the time the court of inquiry considered that the shorter life of the service aircraft was due to their exposure to a multitude of additional stresses when airborne, and this conclusion seems reasonable in retrospect.

Two further steps were taken at this stage. Strain-gauge measurements were made of the stress at the corner of the windows, and these showed that the highest stress, at the edge of the skin near the corner, was in the region of 40 000 psi, about twice the figure which had been calculated by de Havilland, and about two-thirds of the ultimate tensile strength of the material. The figure of 40 000 psi was not in itself very significant but it indicated a high general level of stress in the vicinity. Most of the fatigue cracks in later tests originated at the rivet holes, which cause a localised increase in stress up to a factor of three.

The second step was to recover more of the wreckage, by trawling along the line of flight in the direction of Rome airport. This produced further sections of the wing which showed paint marks, indicating that it had been struck by fragments of the cabin whilst in flight. It also produced a portion of the cabin that had originally been located above the wing, and which contained two windows. By examining the fracture surfaces the investigators were able to pinpoint the probable origin of the failure as a fatigue crack at the corner of one of the windows (Fig. 2.19). When this crack had propagated to a critical length the centre portion of the cabin exploded, the probable lines of separation being as shown in Fig. 2.20.

Further testing

A second series of fatigue tests were made on the fuselage of another Comet I aircraft, using the same tank but applying an internal pressure that alternated from zero to 8¼ psi only.[18] There was no additional loading of the structures. The results of those tests indicate very clearly the way in which fatigue cracks and the final catastrophic failure occurred.

The materials used for the cabin structure are detailed in Table 2.3. The alloy composition was similar to that currently designated as 2024; a high strength copper/magnesium/manganese/silicon alloy, and heat-treated to give two different strength levels. The higher tensile material was used for the skin and the lower tensile for circumferential frames and window frames.

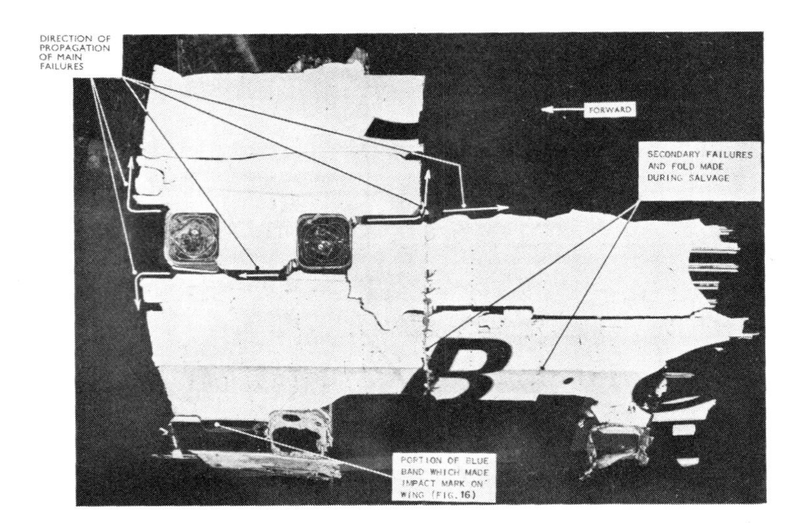

2.19 Wreckage of part of cabin of Comet G-ALYP, pieced together to show the direction of fracture propagation. This section was immediately above the wing.[17]

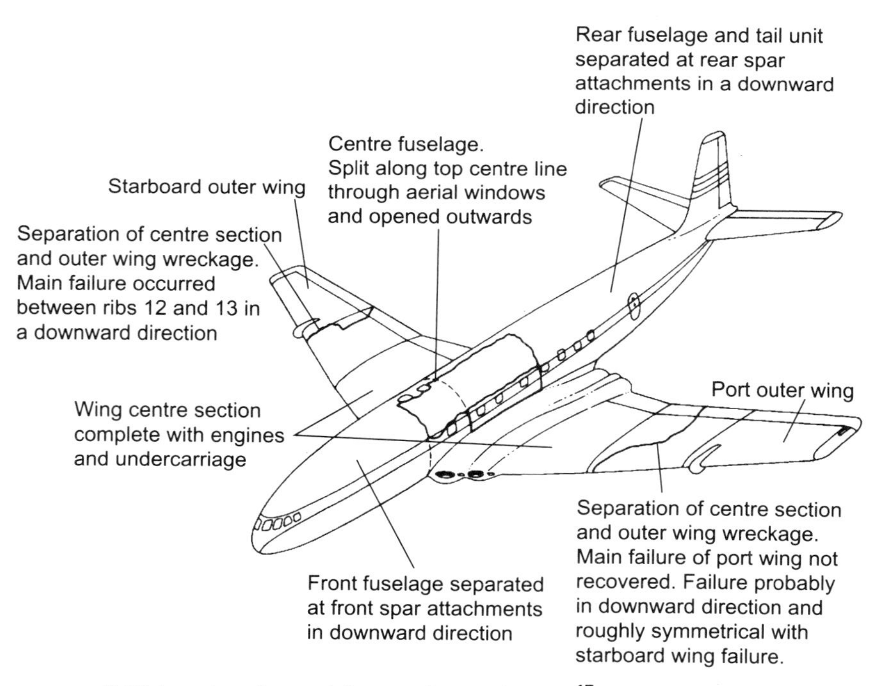

2.20 Location of main failures in Comet G-ALYP.[17]

Table 2.3 Materials used in cabin structure of Comet I aircraft

(a) Composition, percent by mass

Copper	3.5–4.8
Iron	less than 1.0
Silicon	less than 1.5
Magnesium	less than 1.0
Manganese	less than 1.2
Titanium	less than 0.3

(b) Mechanical properties (minimum)

	0.1% proof stress		Ultimate stress		Elongation
	tons/sq in	psi	tons/sq in	psi	(%)
DTD 610	14	31 360	24	53 760	12
DTD 546B	20	44 800	26	58 240	8

Figure 2.21 shows a section through a window frame. The skin in this area was 20 (UK) standard wire gauge DTD 546B (0.036 inches thick, equivalent to 19 US gauge). The window opening was reinforced by a Z-frame with a sealing strip attached to the flange, as shown. All these items were joined, to each other and to the skin, by an epoxy resin adhesive, Redux, and additionally reinforced at the corners, which had a 3 inch radius, by countersunk rivets. Figure 2.22 is a plan view of the rivet holes at the window corner. The de Havilland company was very familiar with the use of adhesive joining for aircraft structures, having used them for military aircraft during the war.

As noted earlier, the stress concentration at the window corners was high, and Fig. 2.23 illustrates a typical case, with a maximum value of nearly 50 000 psi at the edge of the skin.

In the test all fatigue cracking initiated at the countersunk rivet holes in the skin at the window and escape hatch corners. The cracks that eventually propagated catastrophically started at the outer rivet holes. Those that originated at the inner row grew inwards and often terminated at the edge of the window aperture. Such cracks did not become catastrophic. Figure 2.24 shows a typical example, with the crack developing parallel to the long axis of the fuselage (i.e. at right angles to the hoop stress due to the pressurisation) and, in this instance, reaching the critical length for fast fracture at about 6 in. It is characteristic that by the time that the crack was observable, at a length of about ¼ in, 95% of the fatigue life (in terms of numbers of cycles) had already been expended.

The distribution of fatigue cracking along the length of the fuselage is shown in Fig. 2.25, together with the location of the cracks and the number of cycles to failure. This number takes into account the number of

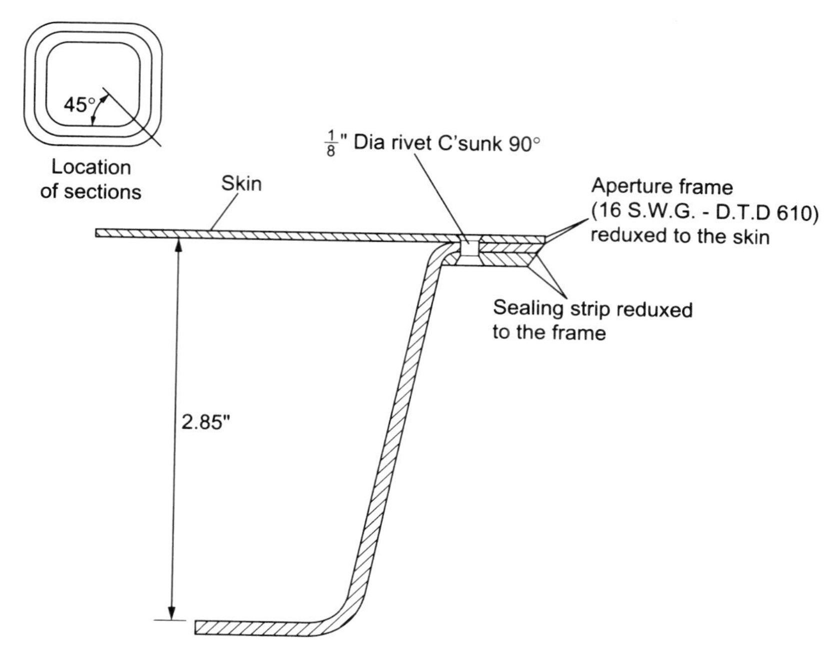

2.21 Section of Comet window frame.[18]

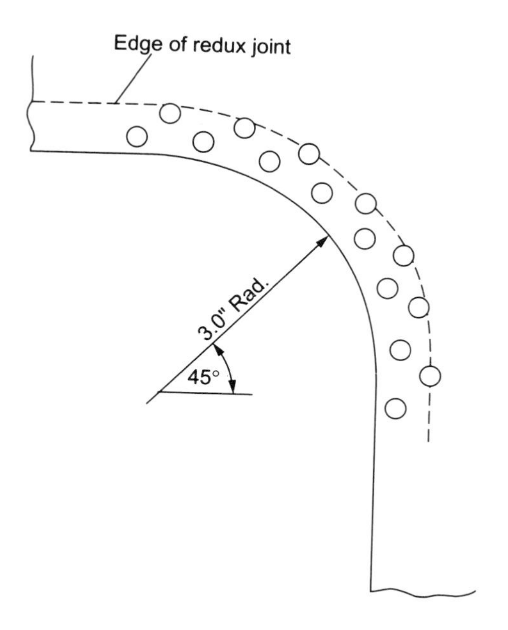

2.22 Detail of Comet window corner.[18]

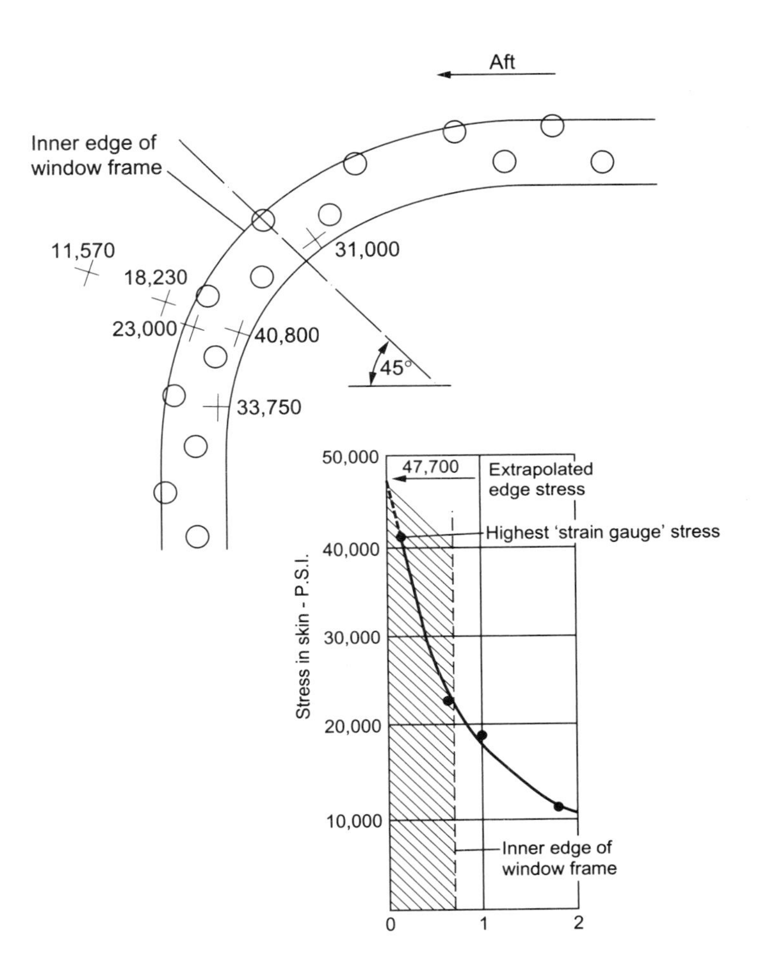

2.23 Stress distribution at the corner window in the Comet aircraft with an internal pressure of 8.25 psi.[18]

pressurisations in service prior to the test. The lowest number of cycles to failure is in the centre section of the cabin, which was the origin of the catastrophic disruption of the Comet I over Elba. The number of cycles to failure was about twice those obtained in the previous test series where a load was applied to the wings, and about six times the number of pressurisations in service before failure. This is taken as a clear indication that whilst the periodic application of internal pressure was the primary

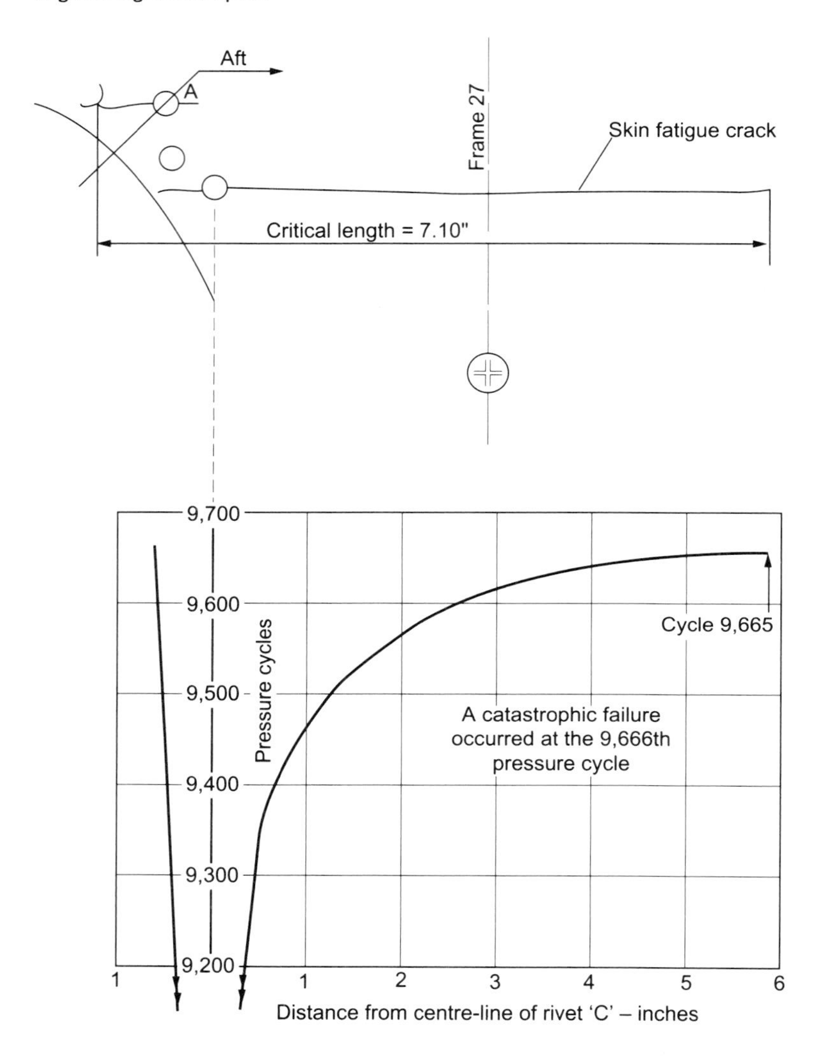

2.24 Growth of fatigue crack in skin of Comet aircraft.[18]

cause of crack formation, service loads may have increased the crack growth rate.

The de Havilland fatigue tests

It will be recalled that de Havilland had carried out fatigue tests on a section of the fore part of the cabin, and that this withstood 18 000

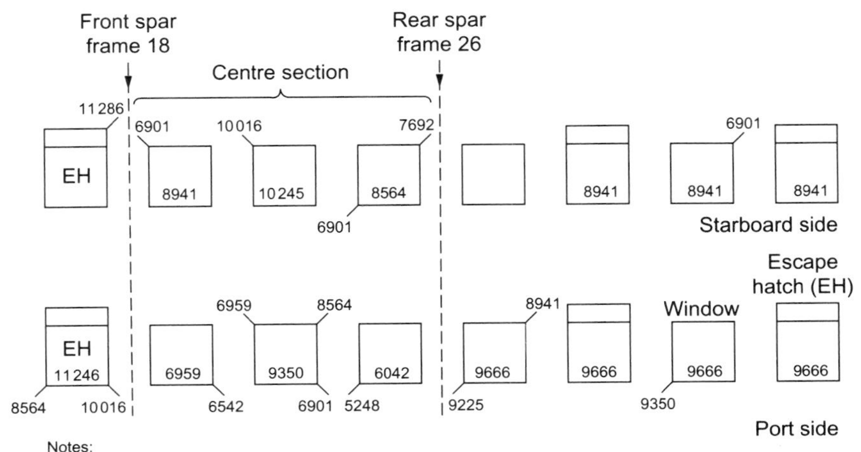

2.25 Distribution of cracks in Comet aircraft fuselage subject to fatigue test.[18]

applications of the working pressure before a fatigue crack appeared. However, the same test section had previously been subject to a number of overpressure tests; 30 applications of between working pressure and twice working pressure. The effect of such overpressure is to cause regions of stress concentration to yield plastically, such that when the pressure is removed they are in a state of compressive stress. When subsequently these same regions are loaded in tension, the strain only becomes tensile after the initial compressive strain has been overcome. Therefore, in a pulsating tensile condition, the tensile part of the loading cycle is reduced. Since it is primarily tensile loading that causes fatigue damage, the fatigue life is increased; in this case about threefold. The beneficial effect of the initial overpressure test is applicable to those areas, in this instance around the outer row of rivet holes, at the corners of the windows, that are most likely to initiate cracks under fatigue loading conditions.

The lesson is that fatigue testing of a structure (or any other test for that matter) should reproduce as precisely as possible the pattern of loading that is to be found in service, and that testing of pre-strained samples is not valid.

The explosion

The Comet I disasters occurred at a time when the discipline of fracture mechanics was under development in the USA, for example at the Naval Research Laboratory in Washington, DC, under the leadership of G R Irvine. A A Wells was seconded to the Naval Research Laboratory during that period and later returned to the British Welding Research Association (now TWI) to conduct some tests on the aluminium allows used for the construction of the Comet.

Fracture mechanics will be discussed in some detail in the next chapter, but for present purposes it is enough to say that a crack in sheet metal will become unstable, and will start to extend at an accelerating rate when the rate of release of strain energy ahead of the crack exceeds the rate of absorption of energy due to the extension of the crack. The longer the crack, the more intense is the strain energy field, and there is a critical crack length, a_c, where it starts to run. The relationship between the critical crack length and the failure stress σ_c is such that ${\sigma_c}^2 a_c$ is constant.

Brittle unstable fracture is commonly associated with steel that is held below a certain temperature. Above this temperature the steel is notch-ductile, below it is notch-brittle. However, high strength aluminium alloys may also behave in a notch-brittle manner. They differ from steel in that there is no transition temperature, and there is little difference in behaviour when the temperature is decreased from room temperature to, say, $-70\,^{\circ}$C. In thick material the fracture surface is at right angles to the plate surface, but in thin sheet (such as that used for the Comet I skin) a shear type failure at 45° to the surface may form. This type of fracture is usually associated with ductile behaviour, but in fast fracture there is little bulk deformation of the type that occurs in, for example, a normal tensile test.

Wells[19] used sheet metal specimens that for the most part were given an internal notch. It was found that for any given material of the value ${\sigma_c}^2 a_c$ was indeed constant within close limits. The material used for the Comet skin, DTD 546 B, was not available but from tests on the same composition but with different heat treatment it was estimated that the critical value of $\sigma^2 a$ for DTD 546B would be 250 inch-(tons/square inch)^2 (equivalent to 39 MN/m^3/$_2$, see Appendix). The stress in the skin of the Comet due to pressurisation alone was about 6.3 tons/square inch so the calculated critical crack length for catastrophic failure would have been

$$250/(6.3)^2 = 6.3 \text{ inches}$$

The actual critical length determined in the second series of tank tests was indeed about 6 inches.

Once an unstable crack initiates it will propagate at very high speed. In brittle materials the crack velocity is up to one-third the speed of sound in the material in question, and in metal speeds of hundreds of feet per second have been measured. The failure of the Comet fuselage therefore could have been, for all practical purposes, instantaneous, and the death of passengers and crew would probably have occurred within a fraction of a second.

The final phase

de Havilland eventually succeeded in producing a sound design and the Comet IV went into service in the late 1950s. It was the first jet aircraft to go into regular passenger service across the Atlantic. But by this time the Boeing Corporation had developed a jet aircraft (the Boeing 707) which had a higher capacity and greater range than the Comet, and this came into use shortly afterwards. So in the end, the Comet was grounded not by technical problems, but because its performance was not good enough.

The Flixborough disaster

Mahoney[20] estimates the capital loss due to the explosion and fire in the caprolactam plant at Flixborough, UK, as being US$161 million at 1992 prices, making it the fifth largest property loss of its type during the 30 years up to 1992. The human cost was relatively much higher; 28 were killed and 36 injured on the site, and outside the works perimeter 53 people were injured and 1821 houses together with 167 shops and factories were damaged to a greater or lesser degree. It was a very grievous accident.[21]

The Flixborough works

The plant was located on the east bank of the river Trent, close to where this river runs into the Humber estuary. It was near the village of Flixborough and 3 miles from the town of Scunthorpe in eastern England.

The site was originally used for the production of fertilisers, but in 1964 Nypro, jointly owned by Dutch State Mines and Fisons Ltd, bought it for the production of caprolactam, which is the raw material for the synthesis of Nylon 6. The caprolactam was at first produced via the hydrogenation of phenol to form cyclohexanone. Then in 1967 the company was restructured and the plant was expanded; at the same time the process was changed so that cyclohexanone was made by the oxidation of cyclohexane.

The unit in which this operation was carried out was considered (only too correctly, in the event) as hazardous, since cyclohexane is a volatile liquid, boiling at about 80 °C, and any leakage could give rise to problems. Nevertheless it was located close to the control room and near to the laboratory and office block. None of these buildings had any protection against explosion and when the disaster occurred the control room was completely flattened, its occupants killed and all records destroyed. The accident took place just before 5 pm on Saturday, 1 June 1974, and the office block was unoccupied. Those in the laboratory managed to escape uninjured, and were able to provide useful information about the place from which the vapour cloud originated.

The process

Figure 2.26 is a simplified process flow diagram of the cyclohexane oxidation unit, and also includes a sketch of the bypass between reactors 4 and 6, and the 8 inch stainless steel line between the two separators S 2538 and S 2539. Both these items were considered as possible sources of the vapour leak.

The oxidation was carried out in a train of six reactors, which were stainless-clad carbon steel vessels set at successively lower levels so that flow between them was by gravity. The temperature was 155 °C and the pressure approximately 8.8 kg/cm^2. Pressure in the reactors was equalised by the offgas line and by keeping the connecting pipes half full of liquid.

Cyclohexane was oxidised by injecting air and catalyst into the reactors. About 6% of the cyclohexane was converted to cyclohexanone and cyclohexanol together with unwanted acidic by-products. The products passed to a series of mixers and separators where the acidic components were first neutralised by caustic soda solution. The products of this reaction were then separated into hydrocarbon and aqueous phases. The hydrocarbon was distilled to separate cyclohexanone and cyclohexanol, which were transferred to another unit for conversion to caprolactam. The residual cyclohexane was returned to the system via a steam-heated exchanger. Make up was provided from a storage tank.

On start-up the system was first pressurised by nitrogen to 4 kg/cm^2, then the cyclohexane was heated to 155 °C, at which point the final pressure of 8.8 kg/cm^2 should have been reached owing to the vapour pressure of the liquid. During start-up pressures sometimes went up to just over 9 kg/cm^2. The relief valves were set at 11 kg/cm^2.

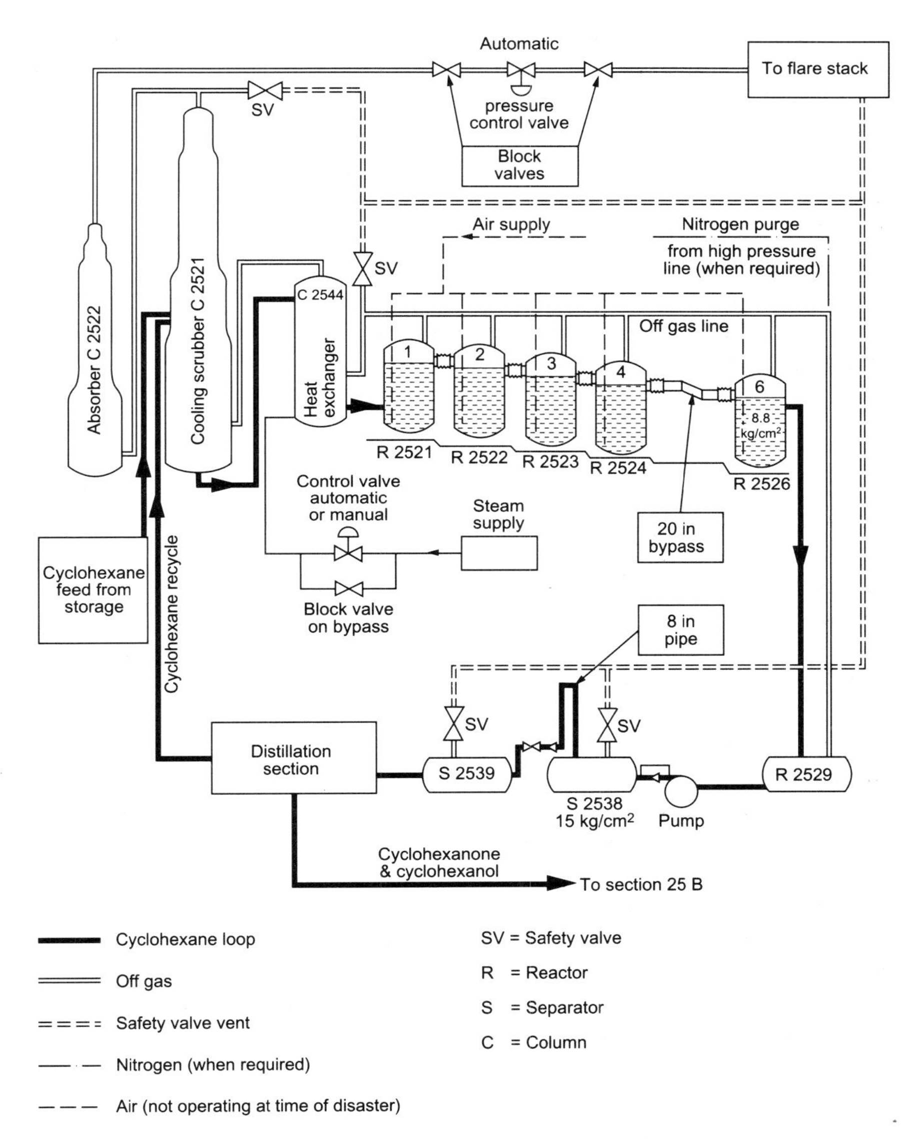

2.26 Simplified flow diagram of the cyclohexane plant at Flixborough.[21]

Removal of reactor no. 5

On 27 March 1974 cyclohexane was found to be leaking from reactor no 5, and the plant was shut down in order to investigate. The following morning it was found that the crack on the outside of the vessel was 6 ft in length. The vessels were fabricated from ½ in thick carbon steel clad with ⅛ in stainless steel. The long crack was in the carbon steel backing material and as a result the stainless cladding was split over a shorter but unknown length. It was decided to remove the reactor for metallurgical examination and to construct a bypass to join reactors 4 and 6. This would enable production to continue.

It will be evident from Fig. 2.26 that the inlet nozzle for reactor 6 was lower than the outlet nozzle of reactor 4, and to accommodate this difference a dog-leg shaped pipe was fabricated with flanges that could be connected to the expansion bellows on the reactors. The original connecting pipes had been 28 inches in diameter but the largest pipe available on site was 20 inches, so this was used. The bypass was supported by scaffolding as shown in Fig. 2.27, with one cross-member under a flange and one under the pipe at each level. After the assembly had been bolted up and the scaffold support poles fixed as shown, a leak test was made with nitrogen at 4 kg/cm^2. A leak was found and repaired, the piping refitted, and a final pneumatic test was applied at 9 kg/cm^2. Construction codes for piping usually specify a hydraulic test at 1.3 times the design pressure, which at Nypro would have been just above the safety valve release pressure of 11 kg/cm^2. Such a pressure would have caused

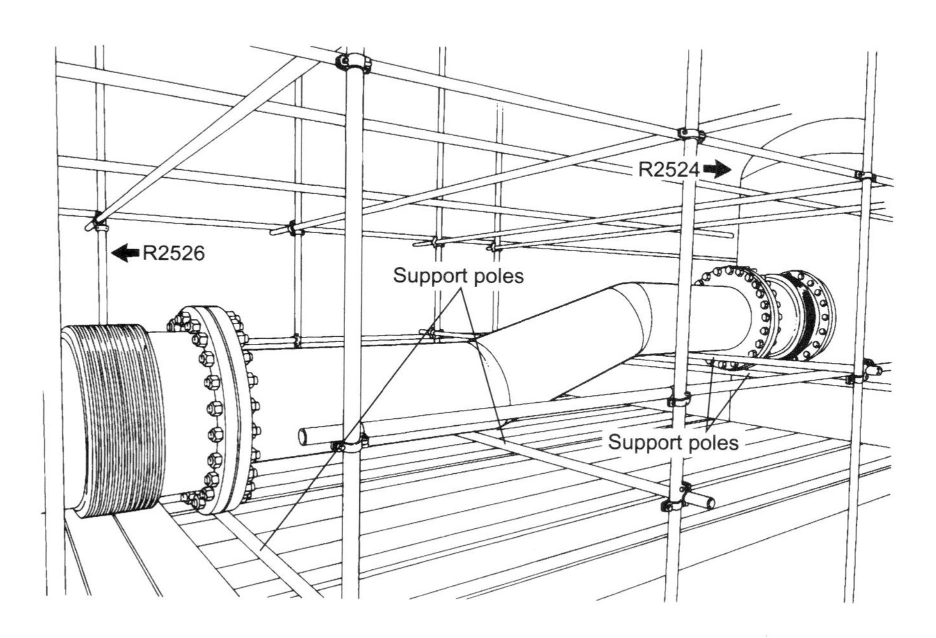

2.27 Probable arrangement of scaffolding supports for bypass assembly at Flixborough.[21]

the bellows to fail and this would have precluded the catastrophe. There was not at the time (nor is there to the present day) any law requiring compliance with details of the piping code, and it is possible that those responsible for the bypass thought that the code would not apply to a temporary structure.

The reason for this fatal weakness, which of course was not realised by any of the Nypro staff at the time, is indicated in Fig. 2.28. Because of the displacement, the assembly as a whole is subject to forces which tend to make it rotate in a clockwise direction. At the same time there is a bending moment acting on the pipe which, if large enough, would cause it to buckle at the mitre joints. The rotating force subjects the bellows to shear loads that they are not equipped to withstand. Bellows are made of relatively thin-walled material (in this instance austenitic stainless steel) and they can bend like a concertina, but if the bending goes too far the convolutions stretch out and the assembly is permanently damaged. Any further increase in pressure is then likely to result in a split or burst.

The works engineer (a qualified mechanical engineer) had recently left the company and had not yet been replaced. No other person on site had

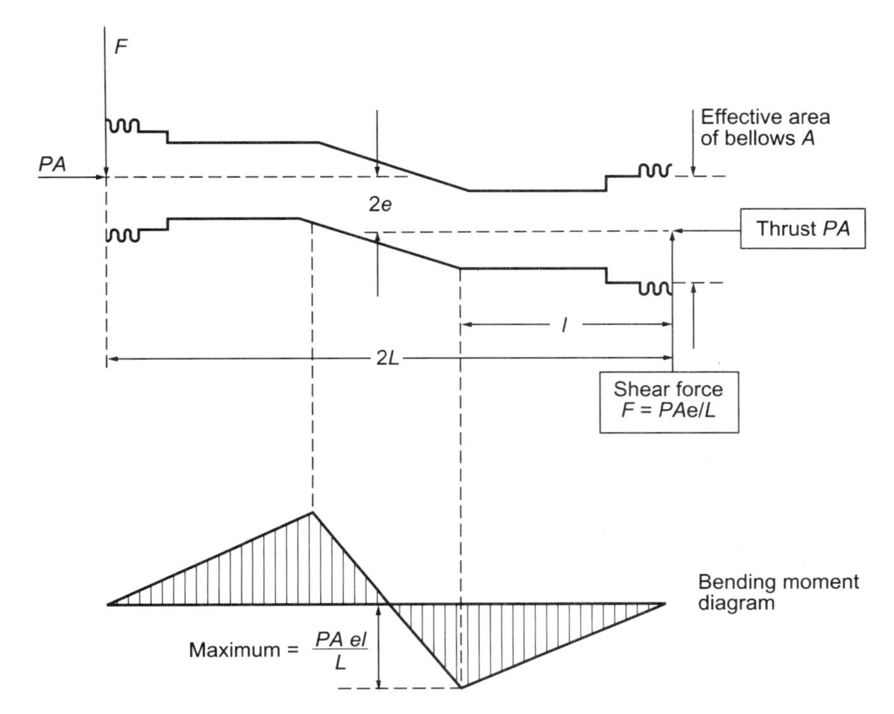

2.28 Shear forces and bonding moment on bypass assembly at Flixborough.[21]

the training or experience to understand these problems or to design piping systems. The 20-inch pipe was checked to ensure that it would withstand the temperature and pressure, otherwise the design work was limited to ensuring that the assembly was a good fit.

Operation with bypass in place

On 1 April start-up operations began and the plant ran normally for about two months. There were no problems with the bypass. There were two short shutdowns during this period but pressure was maintained throughout.

On 29 May a leak was discovered on the bottom isolating valve of a level indicator and the unit was shut down. The leak was repaired and on 1 June circulation of cyclohexane was started again. The start-up went somewhat irregularly owing to further leaks and other problems. The last shift prior to the explosion was from 7 am to 3 pm on 1 June. It would appear that at the end of this shift the full operating temperature and pressure had still not been reached. The operators were required to circulate at the operating temperature and pressure for several hours before starting the air injection, so when the next shift took over at 3 pm normal operating conditions, including oxidation, were unlikely to be achieved for some time.

The explosion took place at 4.53 pm. It was a typical vapour cloud explosion due to a massive release of cyclohexane. Experts estimated that the force generated was equivalent to the explosion of 15–45 tons of TNT. There was property damage over a wide area. Almost all the houses in the nearby village were damaged, and the plant itself was largely destroyed.

The source of the vapour was undoubtedly the 28-inch openings in reactors 4 and 6 which had been formed when the bypass collapsed. The remains of this item were found on the plinth below the reactors. It had jack-knifed, bent through almost 180°. The other possible source of cyclohexane was a burst in the 8-inch stainless steel pipe joining the separators and shown diagrammatically in Fig. 2.26.

Tests on the by-pass assembly

For reasons that were given earlier, the bypass was a most unsatisfactory arrangement, and its collapse was almost certainly the immediate cause of the accident. However, since it had operated for two months without incident, it was considered necessary to determine its actual failure pressure.

Table 2.4 Results of tests on simulated by-pass assembly at Flixborough[17]

Scaffold support No. of poles	Initial position	Temperature (°C)	Test pressure (kg/cm^2)	Result
5	In contact with pipe	155	8.8	No effect
5		155	8.8	No effect
4	Lowered 2/4 inch (6 mm) to represent expansion of reactors	155	8.8	No effect
4		160	9.8	Bellows distorted
4		160	14.6	Bellows burst

To this end a similar piping rig was fabricated and was fitted between reactors 2 and 4, no. 3 having been removed. Scaffold supports were provided as in the original, and the pipe was loaded internally with a chain to represent the weight of liquid. It was heated electrically to 155 °C and pressurised with nitrogen. The reactors were blanked off and filled with water. The results of the relevant tests are summarised in Table 2.4. Collapse of the bellows occurred at 9.8 kg/cm^2, only a little higher than the operating pressure, but they did not burst until the pressure was higher than the relief valve pressure. Other tests reinforced the result. The final conclusion was that under the experimental conditions, jack-knifing and bellows rupture would not occur at or near the normal operating conditions.

The cause of the disaster

Examination of the 8-inch stainless steel line between the separators immediately after the accident showed, in addition to one large and one small rupture, numbers of fine cracks. At the time it seemed possible that these cracks were due to stress-corrosion cracking and that the failure had been a two-stage process; firstly a release of vapour from the 8-inch pipe, followed by an explosion causing the detachment of the bypass, followed in time by the main release of vapour and the explosion that destroyed the plant. However, examination of samples showed that the cracks were due to zinc embrittlement. If molten zinc is in contact with the surface of austenitic stainless steel at a temperature of 800 °C or more, intergranular cracking occurs, and if this happens on the surface of a pipe under pressure, the pipe is likely to burst. At this point it should have been evident that the 8-inch pipe failure was the result of the fire, not the cause of it, particularly as there was no evidence of two successive explosions. Nevertheless, the experts advising Nypro persisted in supporting the two-stage rupture theory, and much of the time of the inquiry was spent in examining, and then finally rejecting, this notion.

There remained the problem of explaining why the by-pass failed at some pressure below 11 kg/cm^2, whereas in the tests a pressure of 14.6 kg/cm^2 was required for bursting. The test set-up was such that as the bellows convolutions blew out and the volume of the mock-up bypass increased, the pressure would tend to fall. Under operating conditions this would not happen; the pressure would remain steady and expansion of gas into the increased volume would result in a sudden input of energy. This would cause the other bellows to fail, which could generate sufficient energy for the pipe to jack-knife, tearing out the remains of the bellows and allowing large volumes of cyclohexane to vaporise. It was calculated that such an event would be likely to occur at a pressure between 10 kg/cm^2 and the relief valve setting of 11 kg/cm^2. Thus, a relatively modest increase in pressure could have triggered the failure.

The other possible cause would be an internal event leading to a sudden rise in pressure. Processes in which hydrocarbons are oxidised by air are notoriously subject to internal explosions. In the cyclohexane unit there was a known risk of explosion if the oxygen content of the off-gas line exceeded 5%, and detectors were in place against this contingency. However, oxidation had not started before the last shift took over, so this type of explosion is improbable. Other possibilities were explored by the Court of Inquiry but were dismissed as likewise improbable. Nevertheless, it must be recalled that the bypass performed satisfactorily, without visible distortion or other distress, for a period of two months before the explosion. During that time it must have been exposed to numerous pressure fluctuations, so that it had, as it were, been proof tested against normal operational variations. A sudden internal event would have provided a much more convincing explanation of the failure than the one described above which was finally adopted by the court of inquiry. Unfortunately, the destruction of all the relevant records makes it impossible to explore this hypothesis other than in a speculative way.

The Flixborough disaster was the first major plant loss to be caused by a vapour cloud explosion following a full-bore rupture of piping. Such failures, as noted in Chapter 1, now result in a substantial proportion of vapour cloud explosions, which in turn have become the predominant cause of large hydrocarbon plant catastrophes. At the time the Flixborough failure was considered to be a direct result of incorrect design of the piping bypass, and this was in turn due to lack of engineering expertise on the site. This judgement needs to be modified in the light of later experience, which shows that large-scale vapour release can occur because of the rupture of piping that has been correctly designed in accordance with the relevant piping code.

The other factor that must be borne in mind is the nature of the process fluid. The general rise in the incidence of vapour cloud explosion has been, in part, due to the increased use of gaseous hydrocarbons and hydrogen. Cyclohexane at 155 °C and 8.8 kg/cm^2 is, if anything, worse since reducing the pressure to one atmosphere will cause boiling and the production of a large mass of vapour very rapidly.

The inherent risk associated with the caprolactam plant was therefore high, although at the time experience had not yet indicated the serious nature of that risk.

Generalisations

It is not, of course, possible, to formulate any objective generalisations relating to catastrophes by looking at the histories of some randomly-selected examples. There are some common features, however, that are worth exploring.

Communication

It is often said that failures of communication are a potent cause of accident in human activities. The loss of the *Titanic* is sometimes cited as a case in point, and of course this may be justified on the grounds that if the officers of the *Californian* had identified the rockets they saw as disaster signals, and if they had roused the radio operator, then they would have known about the collision with the iceberg and could have steamed to the rescue. Whether the *Californian* would have arrived in time is another matter.

In fact the history of that night is rather one of the failure of human perception. Firstly, of course, was the failure to spot the iceberg. However explicable or excusable, this was nevertheless a failure. Secondly, passengers and crew on the sinking ship saw the masthead lights of a steamer, and this steamer appeared to be coming towards them. It was this observation that caused Boxhall, quartermaster of the watch on the *Titanic* to fire distress rockets. But the steamer never appeared nor was it identified subsequently.

On the *Californian* a vessel, thought to be a small or medium-sized tramp steamer, was seen around midnight at a distance of about 5 miles. The captain was concerned to keep his distance so the ship was kept under observation. The second officer of the *Californian*, Mr Stone, saw the rocket flashes and assumed that they came from this second vessel but they did not appear to be distress rockets and eventually the ship steamed off. Again, this vessel was never identified.

The picture is a confused one. The best interpretation seems to be that the *Californian* and the *Titanic* were not in sight of each other. The flashes seen from the *Californian* were in fact Boxhall's rockets. Their explosions were not heard on the *Californian* either because of distance or as a result of some atmospheric freak.

The Piper Alpha disaster has also been categorised as being due to a failure of communication. The day-shift operators did not, it is thought, tell the night-shift that a safety valve had been removed from one of the compressors, and the paperwork gave no indication of this fact. Consequently it was possible, simply by signing a piece of paper, for one of the night-shift operators to pressurise a compressor that was in an unsafe condition. In addition to the failure to communicate between individuals the control system was evidently inadequate. There is much merit in the physical identification of out-of-service equipment on the piece of equipment itself (flagging) as specified by the Minerals Management Service of the US Department of the Interior. It is proposed to extend this system to control panels, pumps and flow control valves upstream of such equipment. A direct means of communication, such as flagging, is more likely to be effective than an indirect system.

Human error figures largely as a cause of failure in the record of aircraft accidents. In many cases, however, some deficiency in the control system may play an important part. The Comet that broke up in a storm after taking off from Calcutta airport was a case in point: the rudder control had no 'feel' so that the pilot may have inadvertently applied excessive force to it, resulting in a structural failure. Another example was the loss of an aircraft at Stockton, UK, in 1967. This crash was caused by engine failure, which in turn was due to leaving a valve in a crossfeed fuel line partly open. This valve was controlled by a lever, where the correct position was indicated by a detent, or spring-loaded notch. The detent had become worn and once again the 'feel' of the control was lost. On the other hand, in an aircraft crash that occurred in 1989, one of the engines of a twin-engined jet started to fail. The crew intended to shut down the defective engine but instead, by a serious error of judgement, shut down the good engine. This was indubitably a case of human error.

Fatigue cracking

The two Comet aircraft losses in the Mediterranean were entirely due to errors in design and in no way to pilot error. In this, and in other respects, the character of this failure was remarkably similar to that which led to the loss of the *Alexander L Kielland* 26 years later. Both were due to unstable fast fractures initiated by a fatigue crack, and in both cases the fatigue crack was initiated by a point of stress concentration at the edge of a circular opening. In the case of the *Alexander L Kielland* the partial failure of a piece of plate under transverse loading, which caused disbonding and lack of reinforcement around the opening made a major contribution to the failure; other similar design features, where there was no disbonding, had behaved satisfactorily at least up to the time of the disaster. The metal used to construct the Comet was not in any way defective. However, it does seem likely that a failure of the adhesive bonding between the aircraft skin and the window reinforcement may have occurred at some stage; in at least one of the post-crash tests the skin rode over a rivet head.

In neither case was any fatigue analysis carried out in the initial stages of design, it being thought that a safety factor used in establishing the design stress would avoid fatigue problems. In neither case was there any redundancy in the design, so that the initial failure caused a catastrophic disruption of the structure as a whole. In both cases the designers were venturing into a new field where previous experience was very limited.

It would be unreasonable to suggest that if the Pentagone designers (or Lloyd's Register, which also participated) had studied the Comet failures, that the *Alexander L Kielland* tragedy would have been averted. Nevertheless, a study of both these failures would be of benefit to all engineers who might be concerned with the design of structures subject to fatigue loading. Even where there is previous experience with a particular type of structure, it must be remembered that apparently minor changes in design can result in the initiation of a fatigue crack.

Vapour cloud explosions

Two of the catastrophes discussed in the chapter were due to vapour cloud explosions. Some of the factors that may contribute to such explosions have already been explored in Chapter 1, but the Piper Alpha and the Flixborough disasters underline one common feature: namely the handling of potentially explosive substances. Gaseous hydrocarbons and hydrogen fall into this category as a matter of course. Cyclohexane would normally be regarded as a volatile and inflammable liquid, in the same category as gasoline. However, at high temperature and pressure the hazard of handling this substance is of a different order; the explosion risk is if anything greater than with gaseous hydrocarbons because of the large mass of vapour that can be generated by the failure of equipment, as demonstrated at Flixborough. Gas leak detectors and dispersion systems offer no defence against a full-bore rupture in plant handling such substances because – as on Piper Alpha – by the time the operator has noted the alarm, or very shortly afterwards, the vapour cloud will have exploded.

For those concerned with the design and construction of hydrocarbon processing plant there are three important steps that could make a major contribution to safety:

1 Identifying those units or parts of units that are handling partially explosive substances and performing steps 2 and 3.
2 Examine the possible sources of internal explosions, and so far as humanly possible, eliminate them.
3 Examine all possible causes of major rupture of piping and, as far as humanly possible, eliminate them.

References

1 'And yet the band played on', *The Times*, 26 May 1994.
2 Garzke W H, Yoerger D R, Harris S, Dulin R O and Brown D K, 'Deep underwater exploration vehicles – past, present and future', paper presented at the Centennial meeting of the Society of Naval Architects and Marine Engineers, New York City 1992.
3 Ballard R D, *The Discovery of the 'Titanic'*, Madison Press Books, Toronto, 1987.
4 Bone James, 'How fragile steel condemned the "Titanic" on freezing seas', *The Times*, 17 September 1993.
5 Spouge J R, 'The safety of Ro-Ro passenger ferries', *Trans RINA* 1989 **131** 1–12.
6 Spouge J R, 'Passenger ferry safety in the Philippines', *Trans RINA* 1990 **132** 179–188.
7 Hooke N, *Modern Shipping Disasters 1963–1987*, Lloyd's of London Press, 1989.
8 Harris J, 'Boredom and the human factor in accidents', *Mater. World* 1994 **2** 590.
9 Anon, 'The *Alexander L Kielland* accident', *Norwegian Public Reports* Nov. 1981 11 (English translation).
10 Gurney T R, *Fatigue of Welded Structures*, Cambridge University Press, London, 1968.
11 Gurney T R, 'Fatigue design rules for welded steel joints', *Welding Institute Res. Bull.* 1976 **17** 115.
12 Cullen W D, 'The public inquiry into the Piper Alpha disaster' Cm 1310 Her Majesty's Stationery Office, London, 1990.
13 Danenberger E P and Schneider R R, 'Piper Alpha – the US regulatory response', in *Offshore Operations Post Piper Alpha*, Institute of Marine Engineers and RINA, London, 1991.
14 *Offshore Operations Post Piper Alpha*, Institute of Marine Engineers and RINA, London, 1991, p. 193.

15 Anon, 'Piper B platform is installed in North Sea', *Marine Eng. Rev.* 1991, 64.
16 Kennedy J, Linzi P and Dennis P, 'The safety background to the design of the Tiffany platform', in *Offshore Operations Post Piper Alpha*, Institute of Marine Engineers and RINA, London, 1991 pp. 81–90.
17 Report of the court of inquiry into the accidents to Comet G-ALYP on 10th January 1954 and Comet G-ALYY on 8th April 1954, HM Stationery Office, London, 1955.
18 Atkinson R J, Winkworth W J and Norris G M, 'Behaviour of skin fatigue cracks at the corners of windows in a Comet I fuselage' R & M 3248, HM Stationery Office, London, 1962.
19 Wells A A, 'The conditions for fast fracture in aluminium alloys with particular reference to the Comet failures', BWRA Research Report RB 129, April, 1955.
20 Mahoney, D, *Large Property Damage Losses in the Hydrocarbon-chemical Industries – A Thirty-year Review*, Marsh and McLennan, Chicago, USA, 1992.
21 Anon, *The Flixborough Disaster*, HM Stationery Office, London, 1975.

CHAPTER 3

The technical background

The record has shown that a significant proportion of catastrophic failures are initiated by the unexpected fracture of some part of the structure; the capsizing of the *Alexander L Kielland* rig and the explosive rupture of the Comet aircraft being two examples described in the previous chapter. It will never be possible to eliminate such failures completely, but their incidence can be reduced by a proper understanding of the failure mechanisms in question.

The other problem that is an increasing scourge in hydrocarbon processing, and which continues to affect shipping and offshore operations, is explosions, particularly of hydrocarbon–air mixtures. The character of explosions, and the ways in which they initiate and propagate, is discussed in Chapter 4.

Mechanical failure

Ductility

In any real structure there are inevitably one or more areas of stress concentration, that is regions where, because of the geometrical arrangement, the stress exceeds the general design stress by a substantial factor. If the material of construction is ductile it will yield at such points, relieving the stress and making the structure more resistant to similar and smaller loading in the future; as in the case of the tests on the Comet fuselage. Metals have this essential ability to stretch to a substantial extent without breaking, and this, combined with their high strength, is why they are used in large engineering constructions.

The ductility of metal results from the fact that the interatomic bonds are not directional, and the crystal structure is in most cases very simple. Because of this, the atoms form planar arrays which can move laterally across each other under the action of shear forces. In practice the plane of atoms does not move as a whole; instead, imperfections or gaps in the lattice structure (dislocations) move across the plane, so that the final effect is to shift the array by the width of the imperfection without losing cohesion.

A number of metals, such as aluminium, copper and nickel, have the very simple face-centred cubic crystal structure, and such metals are ductile at all temperatures. Iron, however, has a body-centred cubic structure at temperatures below about 700 °C and as a result may behave in a brittle or partially brittle fashion at low temperatures. Above about 700 °C the structure of iron transforms to face-centred cubic, the phase so formed being known as austenite. Alloying additions of nickel and chromium retard the reverse transformation on cooling and the face-centred cubic lattice is retained. Consequently the austenitic chromium–nickel steels are ductile down to very low temperatures.

Ductile fracture

The tensile strength of metals is commonly tested using a cylindrical bar, thickened at the ends in order to accommodate grips, and pulled in a tensile testing machine. At one time this was a device with a lever arm along which a weight was propelled manually; currently it is an electronically controlled hydraulic machine. Sheet metal and rectangular test bars may also be used, particularly for testing welds. We are presently concerned, however, with the round testpiece, firstly because it has been extensively studied, and secondly because it displays features that are also observed in real-life fractures, including the brittle fracture of steel. A most useful review of the behaviour of metals during tensile testing is to be found in Ref. 1.

When the load is applied to the testpiece it behaves elastically to start with, that is to say, the stress–strain curve is a straight line and if the specimen is unloaded it reverts to its original length. In the case of mild steel elastic behaviour continues until the stress reaches the yield point, when the testpiece stretches suddenly and the load drops. Current generations of mild (or carbon) steel do not necessarily behave in this way, but they did so in the past, which accounts for the designation 'yield point'. Non-ferrous metals and alloy steels do not have a yield point and the onset of plastic yielding is gradual. The stress–strain plot becomes curved and the yield strength is then defined as the stress at which the curve crosses a line drawn parallel to the straight portion and displaced by the amount corresponding to a small value of plastic strain, usually 0.2%.

After yield, the load continues to increase to a maximum, and then falls until the testpiece fractures. The maximum load divided by the original

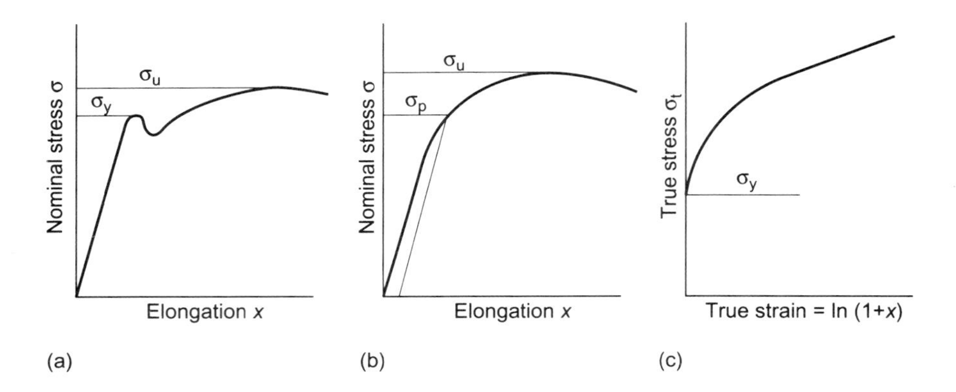

3.1 Typical stress/strain curves obtained on tensile test: (a) with yield point; (b) without yield point; (c) true stress/true strain. σ_y = yield stress; σ_p = 0.2% proof stress; σ_t = load/actual area.

cross-sectional area of the testpiece is the ultimate tensile stress. Typical load–strain curves are shown in Fig. 3.1.

During the plastic extension up to the point of maximum load the bar reduces in cross-section fairly uniformly. After the maximum load, however, a neck forms, usually midway along the reduced thickness, and the cross-section then reduces locally within the neck. Plastic flow is accompanied by an increase in yield strength, owing to a process known as work-hardening. Dividing the load by the minimum cross-sectional area at any specific degree of extension gives the true stress, while the true strain is defined as the natural logarithm of strained length divided by original length. A typical plot of true stress against true strain is shown in Fig. 3.1(c).

At low temperatures and at high rates of loading the neck may not form, and instead failure takes place by 45° shear across the whole cross-section. By contrast, very pure non-ferrous metals may thin uniformly and eventually break when the central section has reduced to zero diameter. Commercially produced metals do not behave in this way. Economy requires that the tensile strength of metals should be as high as reasonably attainable consistent with production cost and ductile behaviour, so that almost all commercial metals are alloyed, and may be additionally strengthened by heat treatment. Almost invariably such strengthening has the effect of increasing the ratio of yield strength to ultimate strength, and of reducing the amount of plastic strain prior to fracture. Table 3.1 lists typical mechanical properties for various materials, and shows how the yield/ultimate ratio increases with tensile strength.

Table 3.1 Typical mechanical properties of materials (in part from Ref. 2)

Material	Elastic modulus (GN/m^2)	Yield stress (MN/m^2)	Ultimate stress (MN/m^2)	Elongation (%)	Yield/ ultimate	Fracture toughness K_{IC} (MN/m$^{3/2}$)	Crack extension energy G_{IC} (J/m^2)[a]	Critical crack opening isplacement δ_c (mm)
Pure metals, annealed								
Aluminium	70	15	40	70	0.4			
Copper	30	33	210	60	0.16			
Iron	210	80	350	40	0.23			
Nickel	200	60	310	40	0.19			
Titanium	120	100	240	80	0.42			
Alloys								
Aluminium 2024 (age-hardened)	72	395	475	10	0.83	70	6.8×10^4	0.17
Cast iron (grey)	100–145		150–400	0	20	3.3×10^3		
Mild steel	210	240	450	15	0.53			
High strength structural steel	210	400	600	20	0.67			
Quenched and tempered $2^1/_4$Cr1Mo	210	525	700	15	0.75	90	3.9×10^4	0.07
Quenched and tempered 4140	210	1700	2000	10	0.85	70	2.3×19^4	0.03
Titanium 6A14V	115	850	950	15	0.9	50	2.2×10^4	0.03
Non-metals								
Concrete	14		45 (in compression)					
Glass	74		50	0	1	0.7	6.6	0.0001
Ice	9		2			0.12	1.6	0.0008
Carbon fibre reinforced plastic	200		1250			60	1.8×10^4	0.01
Glass fibre reinforced plastic	90		1000			90	9×10^4	0.09
Polythene (high density)	0.6		30			3.5	2×10^4	0.7
Timber, along grain	7		115			10	1.4×10^4	0.12

[a] calculated from K_{IC}

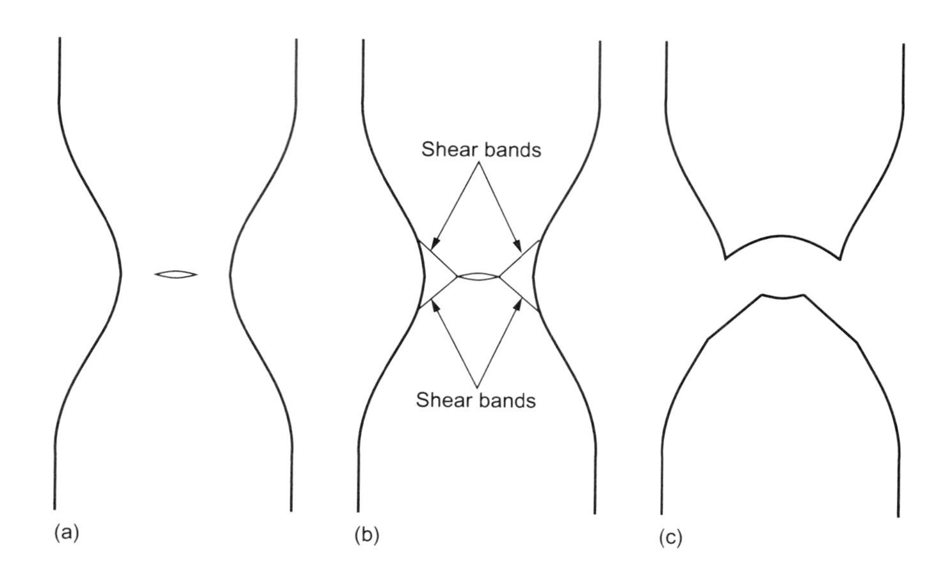

3.2 Stages in the fracture of a round bar: (a) formation of central cavity; (b) 45° slip bands; (c) cup and cone fracture.

The essential features of the normal fracture process of a round tensile bar are shown in Fig. 3.2. Soon after the point of maximum load microcavities form around inclusions in the central part of the necked volume. These cavities result either from fracture of the microinclusions, or by disbonding. The metallic ligaments between the cavities fail in a ductile fashion, and a macrocavity appears. Shear bands then form in the annulus surrounding the central cavity, as shown in the diagram. Eventually microcavities develop in the shear bands, and when the stress is high enough the cavity, which is in effect a crack, extends as a fast catastrophic fracture through the slip band and across the section.

The final result of this process is that one half of the broken testpiece has a cone-shaped tip, whilst the other tip is cup-shaped: a cup-and-cone fracture. The size of the cone and the depth of the cup depend on the degree of elasticity in the testing machine, and if the machine is sufficiently stiff the cup-and-cone may be eliminated. A possible reason for this effect will be discussed later.

The presence of inclusions is not essential to the formation of microcavities in tensile testing. In very clean metals they may form at grain boundaries, at the intersection between grains, or even within grains. In both cases, the process by which microcavities join up to form a macrocavity is called 'micro-void coalescence'.

Most of the ductile extension of a tensile testpiece takes place before the onset of necking, and this, as a fraction or percentage of the original gauge length, provides a measure of ductility. The other common measurement is that of the reduction of area at the neck after fracture. Neither of these quantities represents an absolute measure of ductility since they may be affected by variables such as geometry of the testpiece, rate of straining, temperature and so on. By using a standard testpiece, however, the relative ductility of different samples can be obtained.

The fracture of brittle substances

An ideally brittle material, when loaded in tension at right angles to a crystal plane, will fail by separation of two adjacent planes of atoms, a process known as cleavage. In practice this never happens; fracture is by the extension of pre-existing defects or cracks, and this greatly reduces the potential strength of such materials. The fracture itself is often imperfect, with subsidiary cracking and fragmentation; there may even be some bent fragments, possibly because of local generation of heat.

Brittleness is not an absolute property. Ice is normally brittle but in glaciers and ice-caps it flows and exhibits some plasticity. Rock and glass become plastic when heated to a sufficient temperature. For the most part a material is considered to be brittle if it so behaves in a tensile test at room temperature. Table 3.1 includes some data on the room temperature mechanical properties of non-metals, and it will be seen that even glass has a measurable degree of fracture toughness.

Thermal effects

When a metal is deformed plastically, a substantial proportion of the work done is converted into heat, so that the temperature within a slip band or a fracture surface may increase. Under normal conditions such thermal effects are unimportant during the plastic extension of a tensile test bar, but at the temperature of liquid helium (about $-270\,°C$) the specific heat of metals falls to a low value so that heating and thermal softening along slip planes may be considerable. Consequently, at such temperatures aluminium test bars extend plastically in steps; local heat generation causing a sudden slip which is arrested as the heat dissipates. It has also been shown that in testing alloy steel at $0\,°C$, the intense shear bands that form immediately prior to fracture suffer a temperature rise of $200\,°C$, possibly more. This will cause some degree of thermal softening. It is a matter of common experience that tensile testpieces are hot immediately after fracture.

The way in which such thermal effects can ignite hydrocarbon when a metal container bursts is discussed in a later section on confined explosions.

The origin and nature of non-metallic inclusions

Inclusions of the grosser sort may occur in metals owing to the mixture of slag and liquid metal during refining operations, or to the spalling of refractories, or to some other accident. However, the microinclusions referred to earlier do not occur by accident, but are related to the chemistry of the process.

For example, in the final stage of steel production, metal from the blast furnace is oxidised in order to burn out excess carbon. The liquid metal therefore contains dissolved oxygen. It is then cast, either into an ingot mould or a continuous casting machine, Fig. 3.3. As the liquid metal cools and solidifies, the oxygen solubility falls and consequently oxides precipitate. Sulphides form in a similar manner, the sulphur having

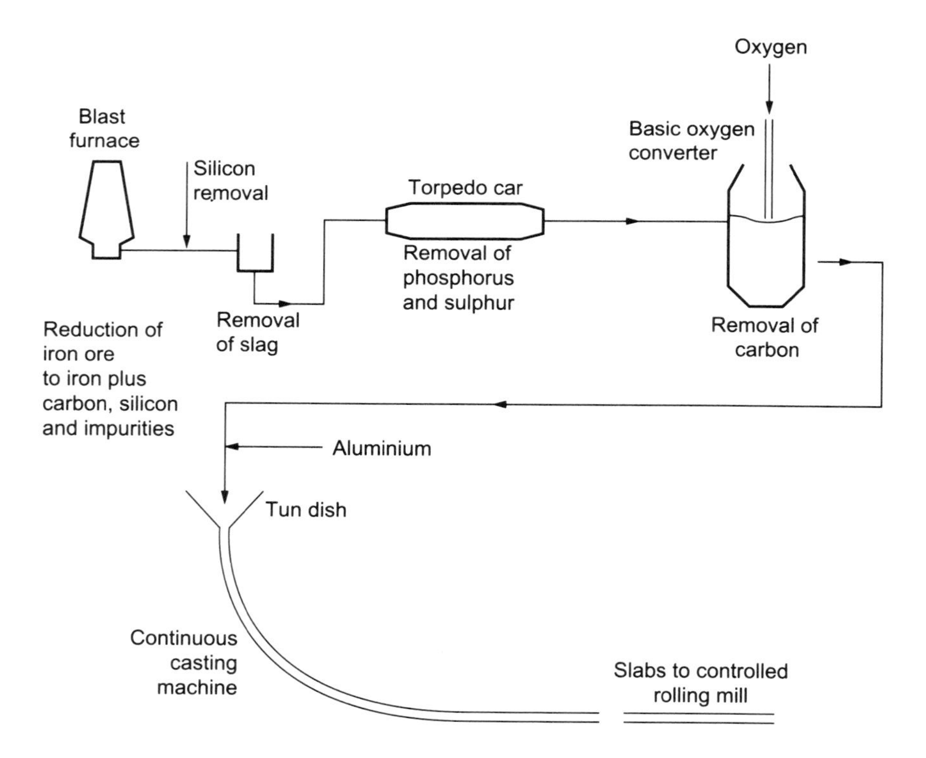

3.3 The indirect process for the production of steel.

been picked up from coke in the blast furnace. Such precipitates are exceedingly small – typically up to one hundredth of a millimetre in diameter, and exceedingly numerous. Similar oxide and silicate precipitates form in weld metal, which goes through a similar oxygen-absorption/oxide precipitation cycle. Figure 3.4 shows microinclusions in stainless steel weld metal. They are larger and more numerous in the sample having the higher oxygen content. As would be expected, a higher oxygen content correlates with a lower impact strength and lower reduction in area in the tensile test.

After the casting operation steel is rolled out to plate or strip. Some of the microinclusions keep their shape, but silicates and sulphides are sufficiently plastic at the rolling temperature to be flattened. One type of

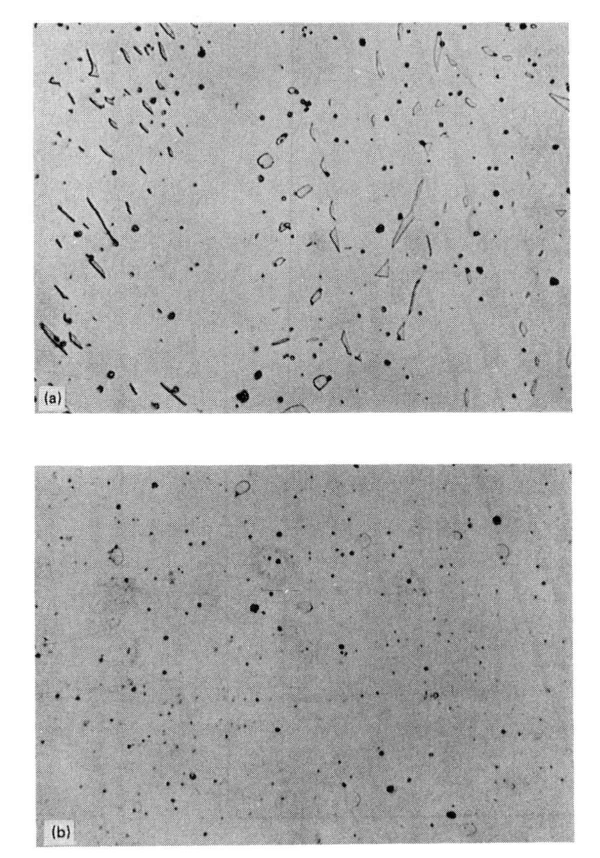

3.4 Microstructure of stainless steel weld metal showing spherical oxide inclusions as black spots: (a) oxygen content 0.12%; (b) oxygen content 0.06% (× 1000).

manganese sulphide is very malleable and inclusions of this type roll out to a pancake shape. When such flattened inclusions are present the transverse ductility of plate material may be seriously reduced.

Non-ferrous metals that are cast, cast and forged, or cast and rolled also contain oxide microinclusions and may be subject to failure by microvoid coalescence.

Failures associated with transverse weakness are discussed in a later section. Before doing so, it will be necessary to describe how the presence of a crack affects the fracture process.

Crack and crack propagation

Consider the crack illustrated in Fig. 3.5. This is a very simple configuration but none the less it is not uncommon in practice. Centre-

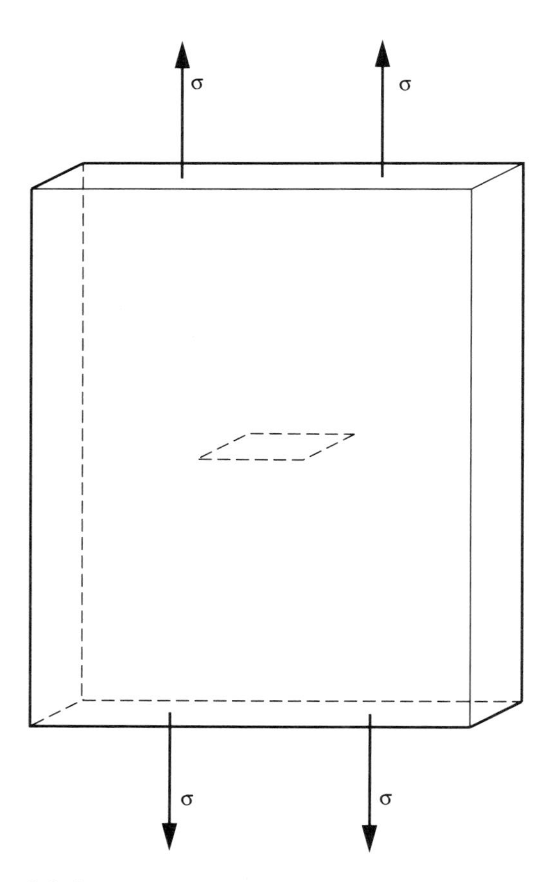

3.5 Centre-cracked plate subject to uniform tension.

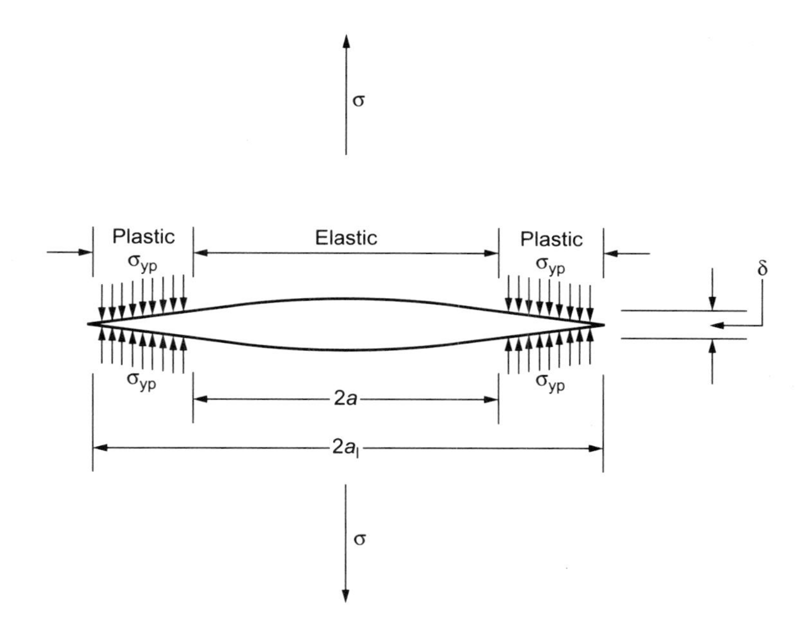

3.6 Detailed model of crack (σ_{yp} = yield strength)

cracked plates were used by Wells to test the material used for the early Comet aircraft, as described in the previous chapter.

The assumed geometry of the crack is shown in Fig. 3.6. Its width is $2a$ and at each end is a wedge of plastically-deformed metal whose maximum thickness is δ. Remote from the crack the plate is subject to a uniform tensile stress σ. If the crack extends at each end, then the rate of release of strain energy is $\pi a \sigma^2/E$, where E is the elastic modulus. The rate at which energy is absorbed by the crack (assuming that plastic flow is confined to the crack tip region) is $2\ \sigma_y \delta$. Therefore the crack will start to run when

$$\pi a \sigma^2/E > \sigma_y \delta \qquad [3.1]$$

The dimensions of the quantity on the left hand side of equation 3.1 are energy per unit area; for example joules per square metre. It is represented by the symbol G and is sometimes called 'the crack extension force'. Consistent with its dimensions, it will here be designated 'crack extension energy'. As the crack length increases so does the crack extension energy, whereas $\sigma_y \delta$ remains the same. Equation 3.1 therefore represents an instability; once cracking starts it will accelerate until the velocity is limited by some other factor. At the point of instability

$$\pi a \sigma^2 / E = \sigma_y \delta_e = G_{IC} \qquad [3.2]$$

where δ is the critical crack opening displacement. For ductile material this quantity may be measured by means of a relatively simple bend test, and is much employed for control of fracture toughness in North Sea offshore structures.

The fracture toughness K_{IC} in the case of a centre-cracked plate is

$$K_{IC} = (\pi a)^{1/2} \sigma_C \qquad [3.3]$$

so that

$$G_{IC} = (K_{IC})^2 / E \qquad [3.4]$$

The stress intensity factor is K_I, arising from an analysis of the stress field surrounding the tip of a crack in an elastic material. The higher the value of K_I, the higher the value of stress at any given radius from the crack tip. The stresses associated with a crack diminish as the square root of the distance from the tip, so that K_I has the dimension force per (length)$^{3/2}$. The concept of stress intensity at the tip of a crack was developed in the USA, and the units of K_I were first expressed as ksi ∇in (thousands of pounds force per square inch times the square root of inches). In SI units this is conveniently translated as Mega Newtons per square metre times square root of metres (MN/m$^{3/2}$). 1 ksi ∇in = 1.1 MN/m$^{3/2}$ so that values of stress intensity or fracture toughness expressed in either set of units will be numerically similar.

The stress intensity factor varies not only with the depth of the crack but also with its shape and orientation, and in general

$$K_1 = k_I (\pi a)^{1/2} \sigma \qquad [3.5]$$

Values of K_I have been calculated for most types of crack.

It will be seen therefore that there are three different quantities, G_{IC}, δ_C and K_{IC}, that may be used to determine how susceptible a material is to fast unstable fracture in the presence of a crack. These three quantities are related to each other by simple formulae, and given one measurement the other two may be obtained. The values of G_{IC} and δ_C listed in Table 3.1 have been calculated from K_{IC}. In all cases, including that of glass, the crack extension energy is much greater than the surface energy of the solid, which supports the view that even in brittle materials, the fracture process is more than just cleavage.

The crack extension energy G_{IC} is clearly a material property on a par with ultimate tensile strength or impact strength. The crack opening displacement is less recognisable as a property, but in the standard test it is possible to define a lower limit value (commonly 0.2 mm, 1/125 in, in the case of carbon manganese steel) below which the material would be regarded as notch-brittle. The fracture toughness is a measure of the intensity of the stress field around the tip of the crack at the point where it is about to run, and is a scaling factor rather than a material property. Nevertheless it allows a quantitative assessment to be made of the effect of different types of crack, and it has additive properties; thus, the stress intensity factor due to a localised residual stress may legitimately be added to that appropriate to a general tensile stress in the member.

The crack extension energy is measurable using a centre-cracked specimen such as that sketched in Fig. 3.5, although the results are normally reported as K_{IC}. One form of the centre-cracked testpiece is the wide plate test, which employs a large specimen and is useful for confirming (or otherwise) the results of small-scale tests. The crack (or crack tip) opening displacement test specimen is shown in Fig. 3.7, whilst that used for the direct measurement of fracture toughness is illustrated in Fig. 3.8. The last two have a machined notch which is extended by a short distance in a fatigue loading machine in order to simulate a sharp crack.

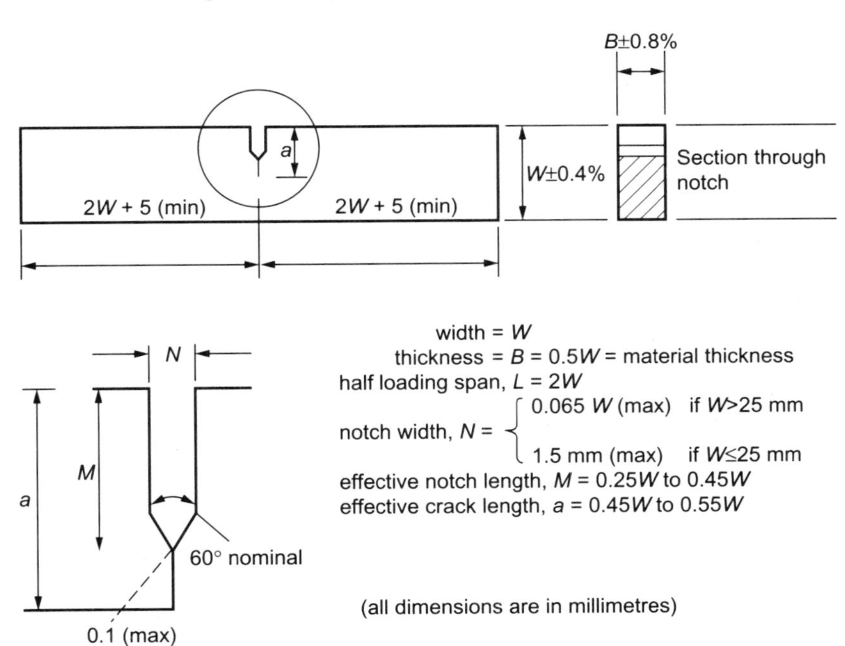

3.7 Crack tip opening displacement testpiece.

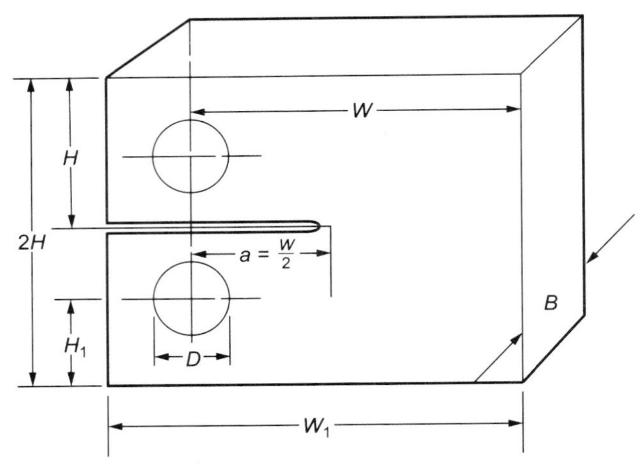

3.8 ASTM compact tension testpiece for the measurement of fracture toughness.

Strain energy sources

In the discussion so far it has been assumed that the crack extension energy is derived entirely from the elastic energy stored in the plate containing the crack. In practice, there is almost always an additional contribution from the surrounding structure. For example, when a test specimen is strained in a tensile testing machine, the machine is also subject to an elastic strain. If the machine parts are massive such that the amount of strain is small, then it is said to be 'stiff' and its contribution to the crack extension energy will likewise be small. This is not invariably the case. For example, it was suggested earlier that rupture of a round testpiece occurs when the central crack extends as a catastrophic fast fracture along shear bands to generate the typical cup and cone fracture. However, the size of the cone is decreased if a stiffer tensile machine is used. This may be explained if the central cavity in the tensile testpiece is regarded as a crack which extends when the strain energy in the testpiece plus that in the machine reaches a critical value. The strain energy release rate associated with the crack increases with crack diameter, whilst that due to the machine is constant up to the point of separation of the two halves of the testpiece. Therefore there is an instability. A stiffer machine makes a smaller contribution to the strain energy available, so the onset of fast fracture is delayed until the central cavity has extended further, and the fracture surfaces are correspondingly flatter.

Impact loading provides a more or less instantaneous strain energy source which may be very large, but it may also be ephemeral. Three instances of brittle fracture under impact loading are considered in a later section.

The brittle fracture of metals

Some authorities avoid the use of the term 'brittle fracture' on the grounds that even in steel at low temperature there is always some evidence of ductile behaviour to be found by microscopic examination of the fracture surface. Moreover, there is usually some reduction of thickness close to the fracture. However, it has already been pointed out that even in brittle materials, energy-absorbing processes occur in the course of fracture, and that thermal effects may even introduce some element of plasticity. Therefore running cracks that occur without bulk deformation will here be termed brittle fractures.

Such fractures may take the form of a 45° shear separation, or may run at about 90° to the plate surface, or they may show a combination of shear and flat surfaces (Fig. 3.9). Shear fractures are characteristic of thin sheet and may appear at the edge of thick plate. Flat fractures in steel may show arrow-like or chevron markings. The surface is usually in part rough or fibrous in appearance, but in more notch-brittle material bright patches of cleavage fracture will be present. There may also be laminar splits which cause the surface to be stepped. In service failures secondary damage may take place after the initial fracture.

Fast unstable brittle fracture will only occur in metals if the material on either side of the crack is substantially undeformed. This may the case if (a) the loading is fast enough to inhibit bulk yielding or (b) the metal is so embrittled that separation takes place at stresses below the yield point. These two cases are considered separately below.

Dynamic loading

In general, the yield point of a metal increases when the strain rate is high, such that it is possible, if the strain rate is high enough and particularly if the structure is inherently rigid in shape, for a brittle fracture to be initiated.

This type of problem arose during the construction of a petrochemical plant. The fire-water system for this plant consisted of a buried ring main to which hydrants (i.e. vertical branch pipes) were welded at intervals. Before the trench containing the main was back-filled, the hydrants stood unsupported. On no less than three occasions these were struck by a

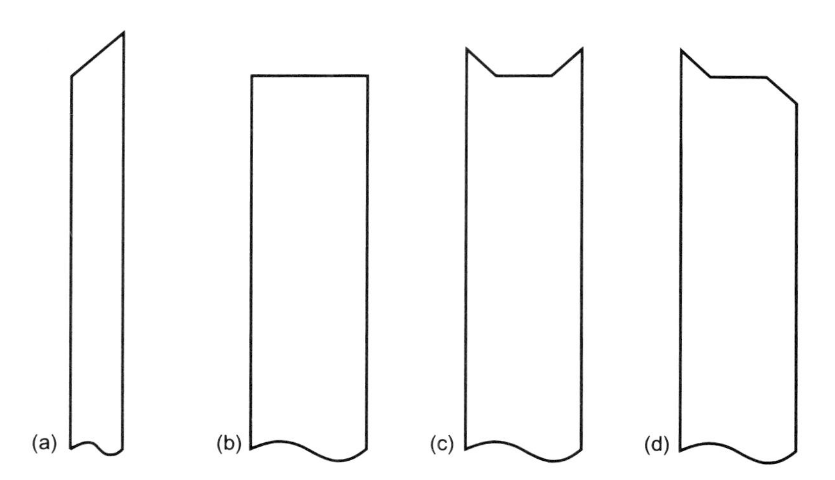

3.9 The shape of various types of brittle fracture: (a) thin material; (b) flat face; (c) and (d) flat face with sheer lips.

vehicle, and on each occasion there was a brittle failure that initiated at the toe of the fillet weld attaching the vertical pipe to the main. It then ran below the fillet around about 90% of the branch pipe circumference, leaving a ligament which bent through nearly 90°.

The material was a pipeline-quality mild steel and the air temperature about 20 °C (68 °F). This was well above the ductile–brittle transition temperature for the steel in question, so that it would seem that dynamic loading had either increased the yield stress, or reduced the fracture toughness, or both.

At the toe of a fillet weld in carbon steel it is common for a small hot tear to form, the depth of which is typically in the region of half a millimetre. Assuming that the brittle fracture was initiated by such a defect, then the stress intensity factor at the toe of the fillet would have been about $\sigma_f (\pi \times 5 \times 10^{-4})^{1/2}$ MN/m$^{3/2}$ where σ_f is the local stress at fracture. Thus, supposing the dynamic fracture toughness of the steel were 50 MN/m$^{3/2}$, then a stress of about 1250 MN/m^2 would be required for fracture. These figures are consistent with the notion of increased yield strength combined with a reduced fracture toughness. It is not practicable to measure the fracture toughness of carbon steel in the notch-ductile condition, but it may be estimated from impact test results to be in the region of 100–200 MN/m$^{3/2}$. If the fracture toughness were not reduced under dynamic loading, the calculated yield stress would be between about 2500 and 3000 MN/m^2, 10–20 times the value measured in a slow strain-rate tensile test.

Brittle fractures were found in welded structural steelwork after the earthquake at Northridge, California.[3] This earthquake occurred on 17 January 1994 in the Los Angeles area; it registered 6.8 on the Richter scale. The motion lasted 10–15 seconds, and the maximum ground accelerations were $12\,m/s^2$ vertically and $18\,m/s^2$, horizontally, very high for an earthquake. The air temperature at the time (4.31 am) was 40 °F, but much of the steelwork was enclosed so that its temperature could have been higher. There was much structural damage, particularly to concrete structures (Fig. 3.10). There were no deaths or injuries however, and none of the structural steelwork collapsed. In more than 100 steel-framed buildings cracks were found at the joint between vertical and horizontal members. Figure 3.11 is a sketch of a typical failure. The crack originated at the root of the weld joining the lower flange of the horizontal girder to the vertical, and propagated for a distance of about twice the weld thickness. In no case was either member severed.

As will be seen from the sketch, the welds from which the cracks originated were made on to a backing bar. Most welding engineers have a poor opinion of such joints, on the grounds that the backing bar may conceal and may even cause weld defects. Therefore, after the earthquake an improved joint was devised. The backing plate was eliminated, the root of the weld was removed and back-welded, and a fillet weld was made between the vertical and horizontal parts in order to reduce the degree of

3.10 Damage to a concrete garage caused by the Northridge earthquake.

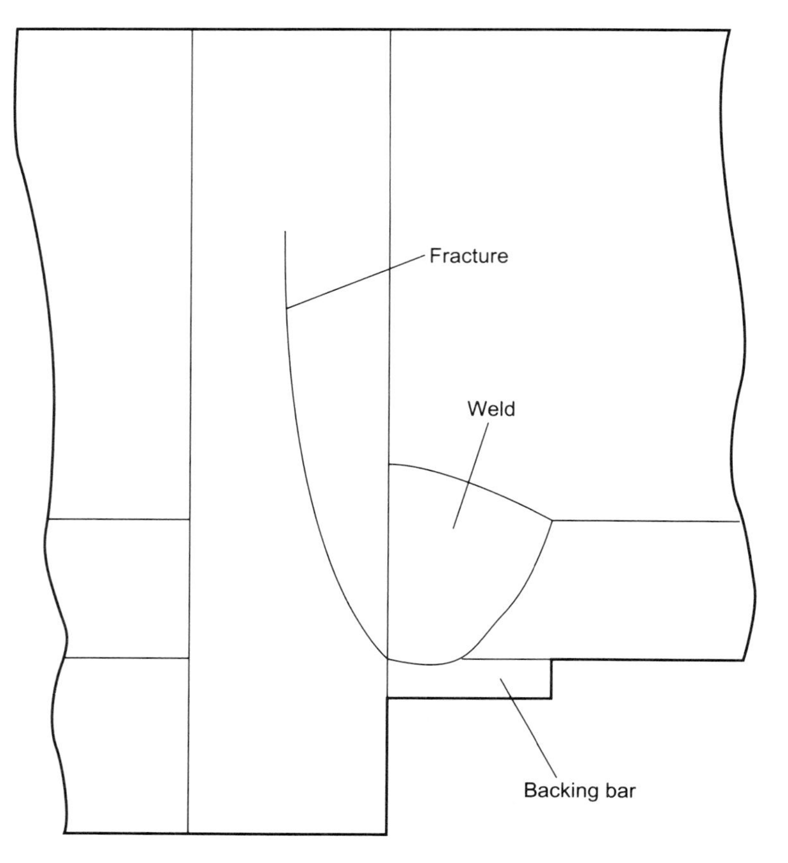

3.11 Typical brittle fracture in structural steelwork after Northridge earthquake.[3]

stress concentration. The improved joints were then tested and were found to be no better than the original type. So much for preconceived ideas about the demerits of backing bars.

It would seem that the Northridge failures had much the same causes as the pipe failure discussed earlier; namely, impact loading acting on an inherently rigid structure with a built-in stress concentration and possibly a small surface crack. In neither case does it seem likely that poor quality material or low temperature played any part.

There is no easy or inexpensive solution to this problem. It is comforting that the cracks at Northridge propagated for a relatively short distance. It has been suggested that this was because the cracks relieved the stress that caused them. It is also possible that they stopped when the earth motion subsided, and that another time the outcome might be less satisfactory.

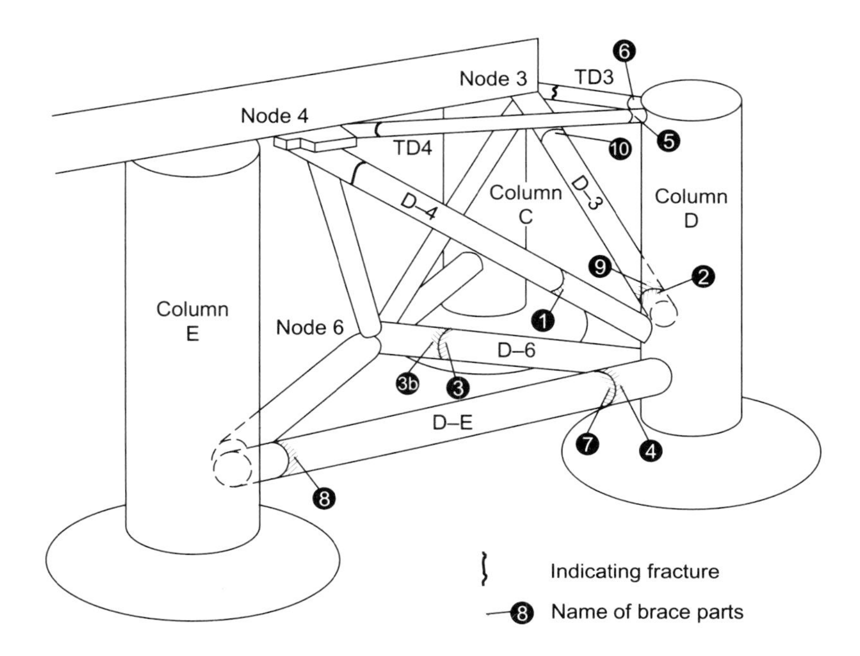

3.12 Fracture of braces around column D of the *Alexander L Kielland* accommodation platform.[4]

The third practical case which is relevant here is the collapse of the *Alexander L Kielland* platform. It will be recalled that following the failure of the D-6 brace, the other braces connecting to column D were subject to a sudden overload, and failed in their turn, causing the column to collapse and the rig to capsize.

The location of the fractures in the various members is shown in Fig. 3.12. It will be evident from this sketch that there was a considerable amount of inertia in the system, to that although the initial brace failure caused a sudden release of strain energy, the transmission of this shock to the other braces would require a finite length of time. The Norwegian engineers calculated that the build-up to maximum load on the remaining members would take 0.2–1.5 seconds. The dynamic effect would nevertheless have amplified the loading to between 1.4 and 2.6 times the normal static load. As a result the braces were stressed to a level above their ultimate strength in tension.

All the members except D6 failed in two places. The sequence was thought to be as follows: firstly D3 and D4 fractured near the column, then D6 also failed and the pontoon started to move outwards, rupturing the ties

TD3 and TD4. At this stage the various members were left projecting from the rig. When it capsized they broke off; an alternative possibility is that the two fractures occurred simultaneously after the first shock.

The fracture surfaces were flat and more or less at right angles to the axis of the tubular member. They were fibrous with arrow markings, similar to those seen on brittle fractures, in a few areas. These were not brittle fractures, however, there being significant bulk deformation. Figure 3.13 is a plot of the contraction at the fracture surface around the circumference of part No 2 of brace D3 (see also Fig. 3.12), which was considered to be one of the initial failures. The contraction varies between about 1 and 10%, and the arrow markings appear to be in regions showing the greatest ductility.

These fractures must therefore be regarded as ductile failures that occurred under shock loading conditions, such that the yield strength was increased to a modest degree and the elongation correspondingly reduced. Bending loads would have caused a maximum stress at one part of the circumference. It is probable that the fracture started here in the same way as in the tensile test bar; namely, by the formation of an internal crack-like void, which then propagated round the pipe in two directions: hence the arrow markings indicated on Fig. 3.13. Arrow or chevron markings are thought to be due to vertical irregularities in the crack front. This front is curved, and the irregularities propagate at right angles to it. Therefore as

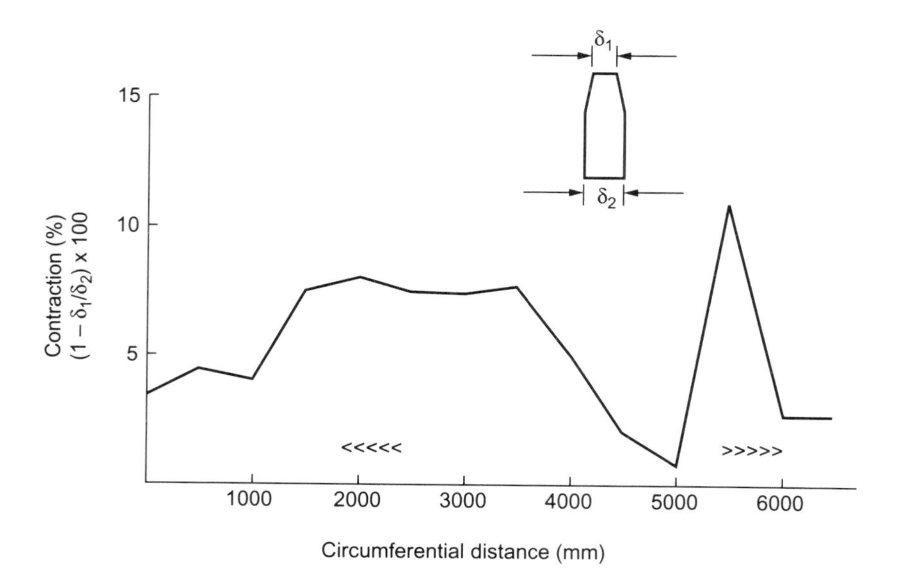

3.13 Contraction at fracture surface: post 2 of brace D3.[4]

the front moves forward they move outward towards the edge of the plate, forming arrows that point back to the origin of the failure.

The circumstances giving rise to the failure of these braces differ quite radically from those discussed earlier. The rate of application of strain was probably much lower, and there was an entire absence of strain concentration effects. The nodes, often regarded as vulnerable areas in tubular structures, remained intact in all cases, and the failures were all in straight tubes.

There were of course similarities; many of the fracture surfaces in the braces were flat and more or less at right angles to the plate surface, and a few showed arrow markings. The essential differences were that in the case of the *Alexander L Kielland* failures, the metal ahead of the running crack was subject to a variable degree of bulk plastic deformation, and that whereas in brittle fracture, relaxation of the elastic strain in the plate is sufficient to cause separation, in ductile tearing a substantial displacement is required to sustain crack propagation.

The embrittlement of steel at low temperature

The transition between notch-ductile behaviour at higher temperatures and notch-brittleness at lower temperatures is measured in the case of steel by making notch-bar impact tests. Figure 3.14 shows a plot of such measurements for samples taken from brace D–E of the *Alexander L Kielland* wreck. This sample was representative of the least notch-ductile of the pieces tested. 'Longitudinal' means that the long axis of the notched bar was parallel to the direction of rolling of the plate, and 'transverse' means that the bar is taken at right angles to the rolling direction. The axis of the pipe was parallel to the rolling direction.

At the upper end of this transition curve the fracture of a notched bar is entirely ductile, and the fracture appearance is 100% fibrous. At the lower end it is partially brittle, showing bright facets due to cleavage fracture. The fracture toughness would, if measurable, show a similar trend. In normal engineering practice, the steel is considered to be notch-brittle when a longitudinal impact energy falls below a certain level. In British and European standards this level is set at 27 J (equivalent to the older standard of 20 ft–lbs) whilst in US practice it is 14 ft–lbs, about 20 J. From the transition curve it is possible to obtain the corresponding transition temperature, which is about −40 °C in both cases. At the time of the accident the sea temperature was about 6 °C.

In the example shown, there is little difference between the transverse and longitudinal transition temperatures, but often the transverse value is significantly higher. Some authorities use transverse impact tests as the

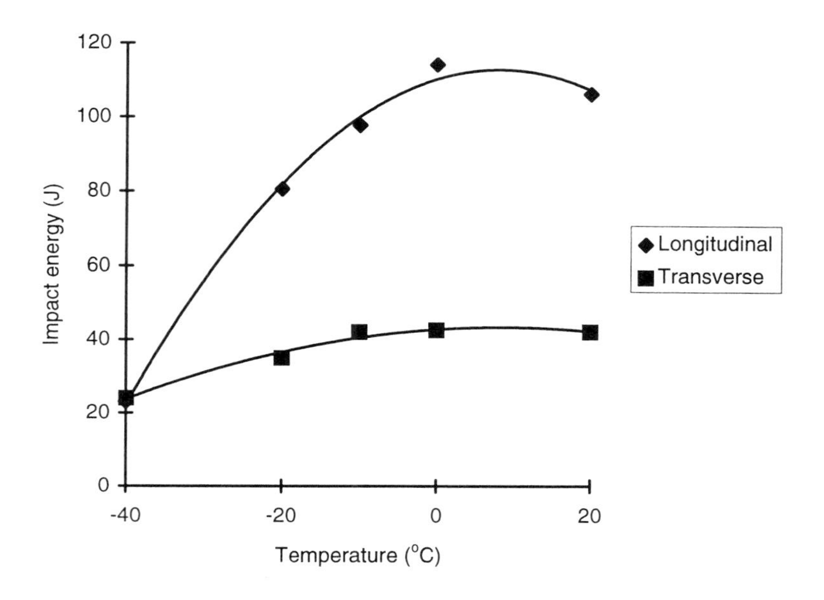

3.14 Charpy impact transition curves for sample 4 from the *Alexander L Kielland.*

criterion, considering this to be safer. However experience indicates that both practices are safe. It is only necessary to specify a higher acceptance level for impact strength if (as in nuclear installations for example) some embrittlement is expected in service.

Improvements in steel quality

It is self-evident that, except where impact loading is the main hazard, the risk of brittle fracture will be reduced by lowering the transition temperature of the material produced by the steelworks. In earlier times when structures and pressure vessels were mostly riveted, there was little incentive to make such improvements. The change came with the move to all-welded construction. Welded steel bridges were erected in Germany and Belgium in the 1930s, and several of these structures suffered brittle failures. Investigation showed that the fractures were initiated by cracks in the heat-affected zone of welds, and that the material around the cracks had been embrittled by strain-ageing. This type of embrittlement results from a combination of plastic strain (due to welding) with modest heating in a steel containing uncombined nitrogen. At the time, nearly all steel produced in continental Europe was made in basic Bessemer converters, where air was blown through iron to burn out the carbon, so that the finished steel had a high nitrogen content. The remedy was to add aluminium that combined with the nitrogen to make it harmless. It turned out that steel made in this way not only resisted strain-ageing, but had a fine grain which produced an elevated yield strength and better impact properties.

Steelmakers in Britain and the USA saw no reason to adopt such measures because in those two countries steel was made in open-hearth furnaces, where the nitrogen pick-up was small. Then, during and immediately after the war came the brittle fracture of the Liberty ships (described later in this chapter), and the need for improved quality became evident. As a start the ratio of manganese content to carbon content was raised and this undoubtedly improved the performance of welded ships. However, there continued to be brittle fractures in pressure vessels, oil storage tanks, bridges and structural steelwork, and in recent years considerable steps have been taken towards making steel that has an inherent resistance to brittle fracture. Much work has been done to remove impurities such as sulphur and phosphorus, and to develop methods of rolling plate and strip so as to produce a fine-grained material. At the same time small additions of niobium, titanium, vanadium and aluminium have been used to make 'microalloyed' steels, which have improved mechanical properties and weldability.

Embrittlement during fabrication

Fusion welding is a very important cause of embrittlement during manufacture. The parent metal adjacent to the fusion boundary is first heated and then cooled, and the cooling rate may be similar to that obtained in the water-quenching of plate. There is therefore the possibility of forming hard, brittle regions. In the case of alloy steel it is normal practice to give the whole welded joint a post-weld heat treatment which softens such hard zones and eliminates any brittleness, but the carbon–manganese and microalloyed steels that are used in structures are not commonly heat treated and here it is necessary that the heat-affected zone should be adequately tough in the as-welded condition.

Strain-age embrittlement has been mentioned earlier. If the steel containing free nitrogen is strained and then treated at about 200 °C, it may be hardened by the precipitation or incipient penetration of nitride. This can be a special hazard if hydrogen cracks form near the weld. The metal at the tip of such cracks is strained, and if this metal is heated to 200 °C by subsequent weld runs, the crack tip is embedded in brittle material. It was this combination of circumstances that caused the brittle failure of bridges, as noted earlier. It is most unlikely to occur now, because most bulk steel is aluminium-killed, and there is no free nitrogen. The root passes of multipass welds may, however, suffer a degree of strain-age embrittlement and although they are unlikely to contain cracks, this may cause problems by reducing the impact strength below specified levels.

A more serious problem is the formation of local brittle zones in the course-grained part of the heat-affected zone adjacent to the weld boundary. Metal in this region is heated to temperatures above 1100 °C and there is some grain growth, which in itself reduces the impact strength. In addition, hard constituents may form within the coarse grains. Some of these are softened by subsequent passes, but some are not. In qualifying welding procedures for North Sea fabrication, it is often required that crack tip opening displacement tests (Fig. 3.7) be made with the tip of the crack in the coarse-grained heat-affected zones. Sometimes the crack tip hits a brittle zone and sometimes not. Consequently there is a wide scatter of results, and it become difficult to set reasonable acceptance standards.

In fact there is no record of brittle failure being initiated in the coarse-grained region of the weld zone, although brittle failures initiated elsewhere have sometimes run along the edge of welds. It must also be recalled that a similar situation exists in the partial transformation region of the heat-affected zone. The microstructure of carbon–manganese steels usually consists of grains of pearlite in a matrix of low-carbon ferrite grains. On heating above about 700 °C the pearlite colonies, which contain about 0.8% carbon, transform to austenite, but the surrounding ferrite does not transform until a somewhat higher temperature. Therefore there is a narrow region of the heat-affected zone which, on heating, consists of islands of high-carbon austenite in a matrix of low-carbon ferrite. On cooling, the austenite transforms to brittle martensite (Fig. 3.15). Again, some of this martensite will be softened by subsequent weld runs, and some will not. So local brittle zones exist elsewhere in the heat-affected zones and these have not caused any problems.

Boiler and pressure vessel codes require that, except for thin carbon steels, welds be given post-weld treatment. Such treatment may have an embrittling effect on the parent plate. It sometimes happens that a completed pressure vessel is modified in design, and areas of the shell are cut out for this purpose. Impact tests carried out on these pieces may give substantially lower results than did the original tests.

Not long ago a heavy carbon–manganese steel pressure vessel was under construction in the field. This vessel was subject to impact testing and testpieces were welded in continuation of the longitudinal seams. The governing code would have allowed these testpieces to have a simulated

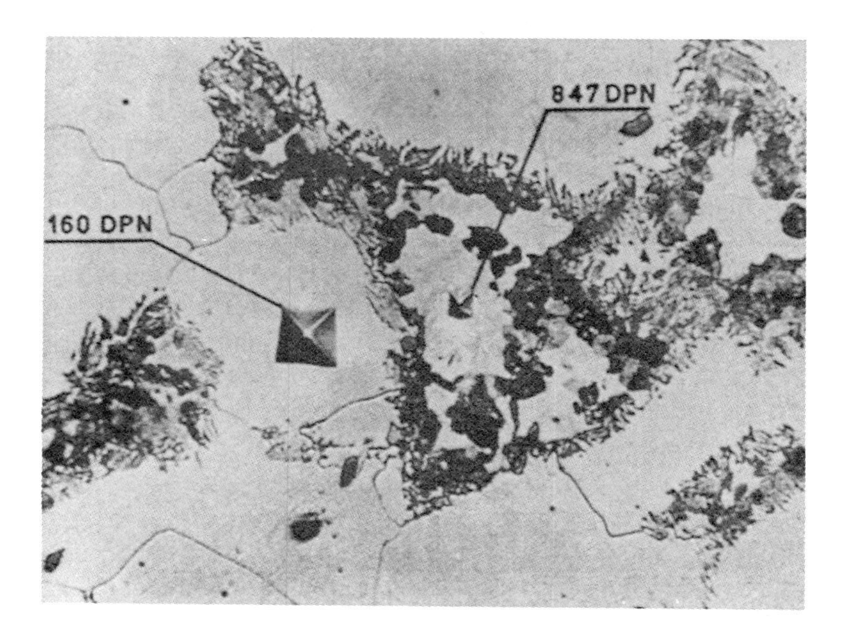

3.15 The partial transformation zone of steel adjacent to a fusion weld. The centre of the pearlite islands have transformed to higher carbon austenite on heating, then to hard martensite on cooling (DPN is the diamond pyramid number, an arbitrary standard for hardness. 847DPN indicates a very high hardness).

post-weld heat treatment in the laboratory, but the client wanted the tests to be fully representative, so that eventually half the testpiece was sent to the laboratory and the other half was heat-treated with the vessel in the stress-relieving furnace. Impact tests made on the laboratory stress-relieved sample passed satisfactorily, but the piece that went through the same heat treatment as the vessel failed. Of course the two heat treatments were not identical; the rate of cooling in the vessel was much lower than that in the laboratory heat treatment. It seems that the embrittlement was due to slow cooling. It is well known that slow cooling of low alloy steel from temperatures of about 600 °C, as used for post-welding stress relief, may result in embrittlement. This defect, known as temper brittleness, was first observed by blacksmiths in the mid-nineteenth century. They found it necessary to quench horseshoes after tempering. Horseshoes are made of a relatively high carbon steel, and temper brittleness (which is discussed further in the next section) is not usually associated with low carbon weldable steels. Nevertheless it is a possible cause of embrittlement in post-weld heat treatment.

Embrittlement in service

Temper brittleness may also result when a low alloy steel is held for a period of time within the temperature range 325–565 °C. It is due to the segregation of arsenic, antimony, tin and phosphorus, which are present as impurities in the steel, to the prior austenite grain boundaries. These are the boundaries that existed in the steel during rolling immediately prior to the transformation to ferrite; the discontinuities that existed at these boundaries persist in spite of the phase transformation. Such segregation predisposes the steel to brittle intergranular fracture. The degree of embrittlement is usually measured by the increase in the ductile–brittle transition temperature, particularly as measured by the appearance of the impact bar fracture. The shift in the transition temperature is increased with larger quantities of impurities and with the presence of certain alloying elements, notably chromium and nickel, and to a lesser extent silicon and manganese. Crude oil containing sulphur corrodes plain carbon-steel at temperatures above about 260 °C so in early crude distillation units handling sour (sulphur-containing) crude, the furnace tubes were made of 5% chromium steel, which resists sulphur attack. However, after a relatively short period of service the tubes became as brittle as cast iron. The problem was solved by the addition of molybdenum, which retards temper embrittlement. Hence the family of chromium–molybdenum steels, 1Cr½Mo, 2¼Cr1Mo, 5Cr½Mo and 9Cr1Mo.

Molybdenum does not completely prevent temper embrittlement, and 2¼Cr1Mo steel, which is used for heavy-wall reactors handling hydrogen at elevated temperature, may suffer temper embrittlement during service. There have been no reports of such reactors failing during operation, but one failure has occurred during heat treatment. The vessel in question was part of a direct desulphurisation unit which was commissioned in 1970, and was one of the first of its kind. It had a wall thickness of 7.6 in, including 0.26 in thick austenitic stainless steel weld cladding. In 1973 it was taken out of use for conversion to a new process, and samples were cut out of the shell for examination. The holes left by this operation were welded up and the affected area, about a quarter of the circumference of the vessel, was being heat-treated when three separate brittle failures took place. These resulted from a combination of thermal stress due to the local heating with temper embrittlement. They were initiated by cracks in the cladding, and these had propagated during service into the 2¼Cr1Mo base metal, which had been quite severely embrittled. It is most unwise to heat-treat a patch in the wall of a pressure vessel; the correct procedure is to heat a complete band including the affected area. Nevertheless, this experience provided a timely warning as to the potential dangers of temper embrittlement in alloy steel. Subsequently, specifications for similar reactors, mostly used in hydrocracking (a process used to break down heavy oils into lighter fractions), have included an accelerated test for the susceptibility to temper brittleness.

Embrittlement due to service at elevated temperature also affects the 25Cr20Ni centrifugally cast reformer furnace tubes, whose failure rate was discussed in Chapter 1. This steel contains a nominal 0.4% carbon, and at the service temperature of about 900 °C, there is a considerable amount of carbide precipitation. As a result the tubes become brittle such that they have little impact strength even at temperatures as high as 1000 °C. Consequently some tubes have suffered brittle failures in service. Such failures are rare because normally the furnace tube assemblies are suspended in such a way that they are not exposed to bending or other external loads. Fractures only happen when there is a (usually inexplicable) failure of the suspension system.

A number of brittle fractures of pressure vessels have occurred during hydrostatic testing. Three steam drums failed in this way in Germany during and just before 1960; two of these were thought to have been embrittled in service. Others happened prior to service. An ammonia converter disintegrated in spectacular fashion in 1965, the fracture was initiated by a crack just a few millimetres long. A steam drum made of the same material, Ducol W30 (a low alloy steel) failed after several previous hydrotests; in this case there was a large initiating crack that had formed during post-weld heat treatment. One storage vessel failed in service with disastrous consequences; this is described later.

Embrittlement and cracking due to hydrogen

Hydrogen dissolves readily in steel at high temperature, for example, in the liquid weld pool during fusion welding. At room temperature or thereabouts it will only dissolve if the gas is very low in oxygen content (generally less than ten parts per million oxygen). Alternatively it will dissolve at room temperature if it is in atomic form. This may happen during corrosion when hydrogen is liberated in the presence of hydrogen sulphide, which has the effect of preventing hydrogen atoms from combining to form H_2 molecules.

Once present in solution it has little effect on the properties of steel except within the temperature range −100 to 200 °C. Within this range the gas segregates to discontinuities; in particular to grain boundaries, to metal/inclusion interfaces and to parts that have been plastically strained, such as the tip of a crack. Where hydrogen concentrates in this way the cohesion of the metal lattice is reduced, and if the region is subject to tensile stress a crack may form and propagate. Because it acts in this way,

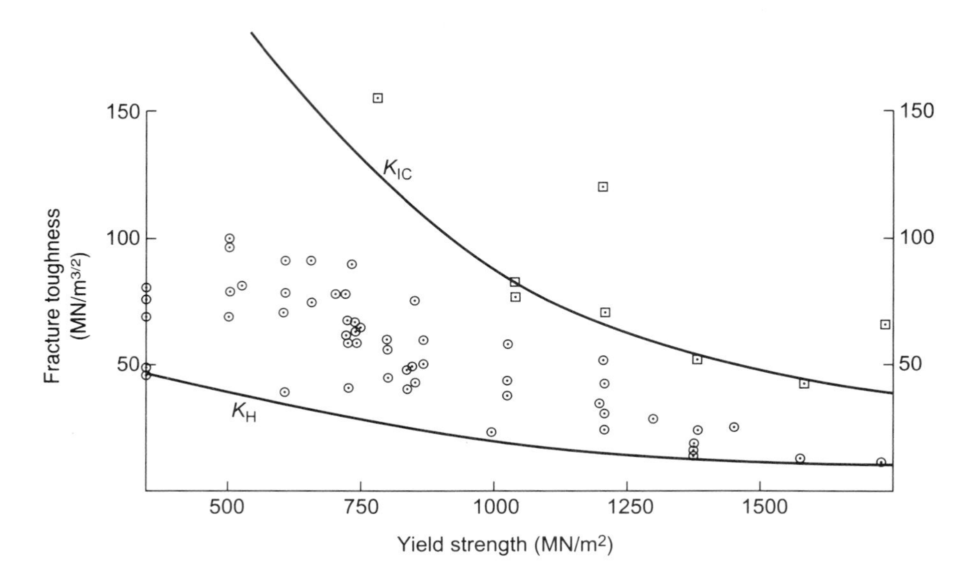

3.16 Fracture toughness of quenched and tempered steels: K_{IC} = hydrogen-free ⊡; K_H = containing dissolved hydrogen ⊙.

hydrogen reduces the fracture toughness of steel, and it is possible to measure this reduced quantity, which is designated K_H. The value of K_H as a function of yield strength is plotted for quenched and tempered steels in Fig. 3.16. Hydrogen cracking is a time-dependent process: a true minimum value of K_H is obtained when the rate of strain, and the corresponding crack growth rate, are low enough to allow hydrogen to diffuse into and saturate the plastic zone at the crack tip. Few investigators have observed this requirement, and therefore the data points scatter widely upwards.

Hydrogen cracking is a familiar characteristic of fusion welds made with damp-coated electrodes or too low a preheat. The cracking usually occurs in the most susceptible parts, i.e. those having the highest yield strength or hardness. In carbon steel this is usually the weld metal, but in alloy steel it is commonly the heat-affected zone. When the stress intensity is low, the cracking is intergranular, but when high it is transgranular. The rate of crack growth is, initially at least, low, such that in welds for example it is easy to follow with the naked eye.

Hydrogen cracking has initiated a number of catastrophic failures, including that of the Kings Bridge in Melbourne and the disruptive failure of the ammonia converter mentioned earlier. The disbonding that led to the *Alexander L Kielland* disaster was almost certainly the result of

3.17 Lamellar tearing below a weld in carbon–manganese steel.

hydrogen-induced cracking, as will be seen in the next section. One of the three instances of catastrophes initiated by a brittle fracture and discussed below was probably the result of hydrogen cracking in a weld. Hydrogen cracking was directly responsible for the explosive disruption of two ammonia converters during service. Details of these two failures have never been published, however.

Lamellar tearing

A macrosection of a weld below which lamellar tearing has occurred is shown in Fig. 3.17. The contraction due to welding has opened up laminations in the steel in a region close to the fusion boundary. These laminations run parallel with the plate surface and are joined by a shear fracture which forms at right angles to the laminations.

The primary cause of this type of failure is the presence of laminar sulphide inclusions. When steel is cast, two main types of sulphide inclusions may form. The first, which appears during an early stage of solidification, is spherical and more or less retains its form when the billet is rolled into plate form. The second type precipitates as intergranular films during cooling. These intergranular films are malleable and roll out to a thin, plate-like form. They reduce the strength and ductility of the plate in a direction at right angles to the surface.

The formation of spherical sulphide inclusions is favoured by the presence of free oxygen, whilst the intergranular type is characteristic of aluminium-killed steel. The formation of non-laminar sulphide inclusions may be promoted in killed steel by the addition of calcium or rare earth metals. Alternatively the sulphur content of steel may be reduced to low levels by calcium treatment of the iron upstream of the oxygen converter, or by treatment of the liquid steel subsequently. By these various means it is possible to improve the through-thickness ductility of steel and thereby improve its resistance to lamellar tearing. Steel grades having a minimum level of through-thickness ductility are available and are used for critical parts of offshore structures.

It is known that the presence of hydrogen in weld metal, because of, for example, the use of cellulose-coated electrodes, increases the risk of lamellar tearing. It may well play an essential role, since the fractures that join the laminations are brittle in character. This may be explained if they are in fact due to hydrogen cracking which is initiated at the tip of the lamination. There is another type of defect that is very similar in its general form, and which is undoubtedly due to hydrogen. This failure mode first appeared in the Persian Gulf during the early 1970s, and affected pipelines carrying moist hydrocarbon gas contaminated with hydrogen sulphide. Corrosion by H_2S results in the release of atomic hydrogen, which supersaturates the metal with hydrogen. The hydrogen initially concentrates at the sulphide–metal interfaces, and then precipitates, forming laminar cavities, which join up as in lamellar tearing and finally lead to complete rupture. Because the steel is supersaturated with hydrogen, the gas in these cavities may be at high temperature. As in lamellar tearing, cracks propagate at right angles to the laminations, joining those discontinuities stepwise. No catastrophes have ensued, but there was considerable disruption and cost when the problem first appeared.

Catastrophes resulting from the brittle failure of steel

The Liberty ships

Two standard types of vessels were built from 1941 onwards in the Kaiser shipyards, the Liberty ship, which carried general cargo, and the T2 oil tankers. Both were fully welded, and for the time the speed of construction was remarkable; the first T2 tankers took on average 149 days to complete but by the tenth round of contracts, this time was down to 41 days. Large numbers were produced, and they made a massive contribution to the war capability of Britain and later the Allies.

At the start there was virtually no experience of welded ships and there were some built-in stress concentrations, notably at square notch corners, the end of bilge keels and at cutouts in the shear strakes. These discontinuities were the origin of most of the serious brittle fractures. The loss rate became serious in the winter and spring of 1942–43 and peaked during the same period of 1943–44. The US ship structures committee later analysed the records and classified the types of failure. Class 1 included those where there were one or more fractures that endangered the ship or resulted in its loss. For the original design the rate of Class 1 casualties was 4.3 per 100 ship–years, as compared with losses from all causes of 1 per 100 ship–years immediately before the war.

After the severe winter of 1943–44 the design was improved by reducing the severity of notches and by introducing a minimum of four riveted seams running the length of the hull and intended as crack arresters. With the improved design details the Class 1 casualty rate fell to 0.5 per 100 ship–years. The riveted crack-arresters were less successful; without them T2 tankers had a Class 1 failure rate of 1.9 per 100 ship–years, with the riveted seams that figure fell to 1.2.

The other factor that became evident in 1943 was that the casualty rate was worse at low temperatures and in rough seas. So from that time the Liberty ships were, wherever possible, routed through calmer and less icy seas.

The impact properties of plate from some of the casualties were measured. It was found that plates in which brittle failures originated had a 21 J transition temperature of 15–65 °C with a mean of 40 °C. Such steel would undoubtedly have been notch-brittle during winter and spring in the North Atlantic. Improved steel qualities were developed soon after World War II and the incidence of brittle fracture fell. Weld quality was likewise improved. Some of the early brittle fractures were initiated by weld defects and particularly by cracks. The improvement in steel quality continues; in 1991 Lloyd's Rules included provision for 16 grades of as-rolled or normalised and 18 grades of quenched and tempered steel. All are impact-tested except the A grade, which is a general-purpose mild steel for use in non-critical locations. The lowest impact-tested grade calls for a 27 J transition temperature of 0 °C, the highest for an impact strength of 41 J at −60 °C.

The Sea Gem[5,6]

This ill-fated vessel started life in the army. It was built in 1953 as a jack-up section of an aerial tramway (cableway) for the US Transportation Corps, and operated as such during 1955–56. It was then mothballed until 1962 when it was declared surplus to US Army requirements. After being purchased by the French contractor Hersent, the hull was modified and it was used as a flat-topped barge in the Persian Gulf, under the name *GEM 103*. GEM was the Compagnie Generale d'Equipments pour les Travaux Maritimes, a joint subsidiary of the De Long Corporation of New York and Hersent of Paris. In the spring of 1964 *GEM 103* was towed to Bordeaux and re-converted into a ten-leg jack-up platform, using De Long jacks (to be described later). For the rest of 1964 it was used as a civil engineering platform.

At this time British Petroleum was assembling material for use on the newly discovered North Sea field and it was decided that *GEM 103* would

3.18 The *Sea Gem.*

be suitable for North Sea drilling subject to certain modifications. So the rig was hired by the contractor Wimpey, who remained responsible for maintenance, and then re-hired by BP.

The original hull was 300 ft in length, 90 ft in beam and 13 ft deep. In order to provide a drilling slot, a length of 100 ft was cut off the stern and a newly fabricated section 47 ft long was added. A heavy duty crane and the drilling derrick were mounted on the deck, together with other necessary equipment such as accommodation, a radio room and a helideck. Two legs were mounted in the new section, making ten in all, as before. The legs, which were 5 ft 11 in in diameter were cut and a new central length of 128 ft inserted, making them 220 ft in length altogether. The new part was intended to be that normally engaged by the jacks. Figure 3.18 shows the rig, now called the *Sea Gem*, after these modifications.

During the construction period there was considerable discussion about the means of avoiding brittle fracture, and in particular about the level of impact energy to be specified for the plate material. Eventually the figure of 25 ft–lb at 0 °C was established as the basic requirements; material not meeting this figure was to be normalised. Normalising consists of heating the steel to a temperature above the ferrite–austenite transition, say 950 °C, and cooling in air. This refines the grain and improves notch-ductility and impact strength. In fact, owing to an error in translating metric units to foot-pounds, the borderline level for normalising was 20 ft–lb at 0 °C.

It is not clear from the record whether any tests were carried out on the material of the original structure, which dated from 1952. Presumably not, because a failure would logically have required that the whole barge be rebuilt.

The De Long type D jacks (Fig. 3.19) consisted of a circular framework mounted on the deck which contained two sets of grippers. These were made of nylon-reinforced neoprene rubber, and were forced against the leg by air pressure. The upper set of grippers was attached to the decks by four equally spaced tiebars. The lower grippers were connected to the upper set by 12 pneumatically operated cylinders (visible in the photograph) which were supplied with air at 350 psi by a set of four compressors. There were non-return valves that retained the pressure on the grippers in the event of a failure in the air supply.

To raise the platform the lower set of grippers remained fixed and the upper set was released. The cylinders were then pressurised, which raised the upper grippers and the platform by 13 in. The upper grippers were then applied and the lower ones released. Retracting the cylinders raised the lower grippers by 13 in so that the cycle could be repeated. To lower the platform the operation was reversed. Except when raising or lowering, both sets of grippers were applied. At all times the platform was suspended from the jacks by tiebars, and the integrity of the structure relied, firstly, on the strength of the tiebars and, secondly, on the frictional force between the grippers and the legs. Operators were provided with carbon tetrachloride to clean off any grease and fine sand to improve the grip.

The *Sea Gem* was towed out to the required position and on 3 June 1965 was successfully jacked up to the drilling position 50 ft above sea-level. The sea-bed was sand over boulder clay; the legs were designed to penetrate a few feet into the sand for better stability, and this they appeared to do. Drilling operations proceeded without undue problems, the well head was secured on 18 December, and preparations for moving the barge to a new location were put in hand. Such a move required at least 48 hours of favourable weather. Forecasts indicated that conditions could be right at the end of December, and as a preliminary step the platform was lowered 12 ft. This was done by operating each jack individually. The

3.19 Type 'D' De Long air jack fitted on drill barge *Sea Gem.*

lower gripper was released, the cylinders extended, and the gripper applied again. When all jacks had functioned in this way, the barge was lowered one stroke by working all jacks together from the central control. The upper grippers were all released, the cylinders depressurised, and the grippers applied again.

The accident

On 27 December the weather conditions were considered to be good enough to lower the platform a further 10 ft. The air temperature was about 3 °C and the water temperature about 6 °C with a force 5 wind and waves not higher than 10 ft. The jacks had proved to be in good order for lowering, but in view of the possible need to raise the platform if conditions deteriorated, the operator decided first of all to raise it by one stroke. This operation was conducted from the central control; the upper grippers were released and when the pressure at the cylinders had reached 300 psi the forward end of the barge started to rise. At the full pressure, however, the aftermost jacks had not moved at all. The operator made a visual check, and then started to release pressure. At this point there was a loud bang and the platform lurched violently to port. This movement was suddenly arrested, probably because the port side legs jammed in their wells. The platform then righted itself and fell more or less horizontally into the sea. For the time being it floated, but the fall had caused a brittle crack to extend right across the bottom, and water was pouring in. The wreckage remained afloat for 10–15 minutes, then capsized and sank.

During this time some of the crew showed commendable presence of mind and courage. Two liferafts were launched, both from the starboard side. Five men got into No 3 liferaft and 14 into No 1. The toolpusher, who is normally in charge of drilling operations and was the senior man on board, took command of No 1 raft. At the time there were ten men on the helideck or thereabouts. Mr Hewitt, the toolpusher, called on these men to come down and board one of the liferafts, but they stayed where they were. When the situation became too dangerous Mr Hewitt cast off. Even then he manœuvred the raft as close to the helideck as possible in an attempt to rescue these men, but was not successful. The liferafts were then paddled or drifted over to the *MV Baltrover*, a ship on passage from Gdynia to Hull, and which had passed within sight of the rig when it collapsed. With some difficulty all those on the liferafts were taken on board this ship. All those who remained on the wreck died, including two who were rescued from the sea by helicopter but died later. During the initial phase of the collapse both the radio room and the *Sea Gem*'s one lifeboat had been pitched into the sea. The radio operator swam over to the lifeboat and got aboard, but died of cold.

The cause of the accident

The inquiry tribunal[6] had few doubts about this question; the most probable cause of the collapse was the brittle failure of tiebars on the port side of the rig. Moreover, most of the fractures of legs and other parts of the structure that occurred as a result of this initial failure were brittle fractures. Figure 3.20 shows the remains of one broken leg. The fracture is completely brittle.

The tiebars or suspension links, however, were the real problem. The form and dimensions of these members are shown in Fig. 3.21. They were flame cut from 2½ in thick plate, normally to ASTM A36 but possibly to an equivalent French standard. A36 is a structural grade of carbon steel

3.20 Fractured caisson of *Sea Gem* 31 January 1966.

with no special requirements for notch-ductility, the only concern of the designers was with its tensile strength. The jacks, including tiebars, were fabricated by the firm of Lecq of Douai, France in early 1964 when *GEM 103* was made. The material was not subject to impact testing or normalising.

There were a number of reasons why the Tribunal considered that fracture of the tiebar was the initial cause of the disaster. These were

1 The divers' survey of the wreckage showed that tiebars on the port side were either missing or fractured.
2 One complete and six fractured bars were recovered, and all the fractures were brittle in character.
3 The temperature at the time of the accident was 3 °C, low enough for the steel to be in a notch-brittle condition.

In addition survivors recalled that immediately after the platform fell,

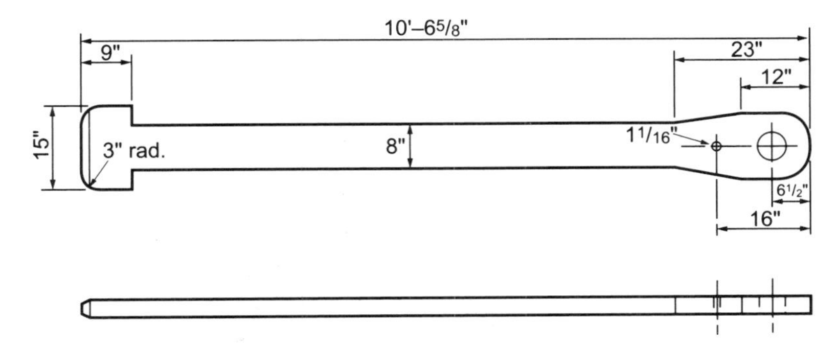

3.21 Dimensions of tiebars of De Long air jacks on *Sea Gem.*

two or possibly three jacks were left attached to the legs.

There had also been a previous failure. On the night of 23/24 November a loud bang was heard, and investigation showed that two tiebars on the No 12 leg had broken. These were replaced with spares within a few hours and there was no concern about the incident among those on board. After the loss of the *Sea Gem* the two links were examined by the UK Safety in Mines Research Establishment who found that the failures were indeed brittle and had initiated at a sharp corner radius at the spade end of the member (later measurements showed this radius to be 0.08 in). In addition, there were gouge marks in the flame cut edges and several weld runs where an attempt had been made to repair such gouges. The hardness of the heat-affected zone of these weld runs was up to 480 Vickers, and averaging 300, as compared with about 140 for the original plate. The welds contained porosity and slag inclusions (Fig. 3.22), and cracks were found near the fusion boundary. One of the brittle failures was identified as having initiated at a weld.

It was also found that the pin holes were slightly elongated, indicating that at some stage the links had been stressed beyond the yield strength. The tiebars recovered from the wreck were examined in the laboratory of Lloyd's Register of Shipping, and these all showed a similar feature. The reason for this apparent overloading has not been explained.[7]

Lloyd's[8] carried out impact tests on samples from the seven pieces of bar, the results have been averaged and are plotted as a transition curve in Fig. 3.23. The transition curve for steel from the *Alexander L Kielland* wreck is plotted on the same diagram to show how the notch-ductility of structural steel had improved in the 15 years between the two catastrophes. The mean impact strength of the tiebar material at 0 °C was 17.5 J, about 13 ft–lb. These items therefore combined low notch-ductility with the presence of serious defects due to surface weld runs which had not been heat-treated. At the time of the accident, conditions

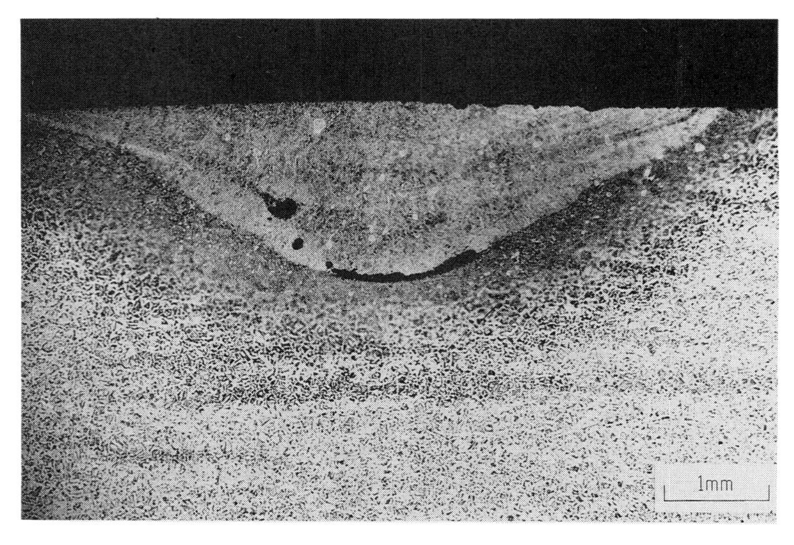

3.22 Section of a typical weld run on one of the *Sea Gem* tiebars.

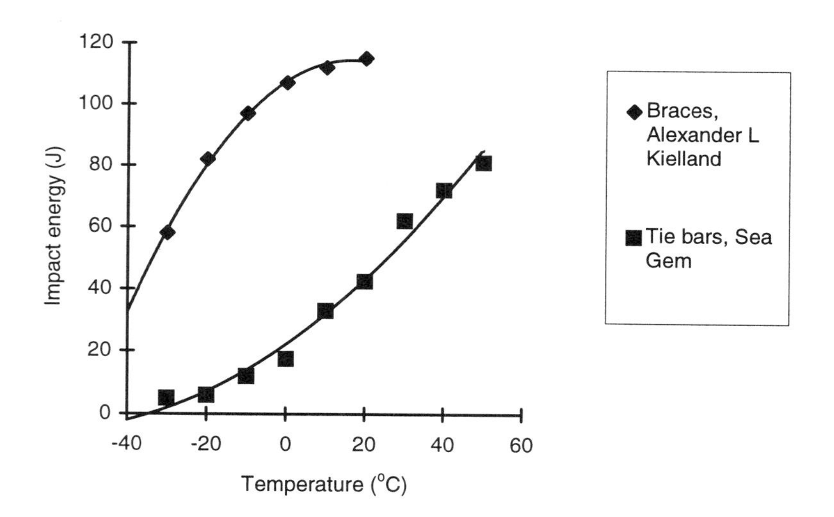

3.23 Charpy V energy transition curves; tiebars from *Sea Gem* compared with samples from braces of *Alexander L Kielland* wreck.

for initiation and propagation of brittle cracks in the tiebars were present in full measure.

Ammonia tank failure[9]

Liquid anhydrous ammonia is used as a fertiliser, and it is stored either at room temperature under pressure or at atmospheric pressure and maintained at subzero temperature. At a fertiliser plant in Potchefstroom, South Africa, the ammonia was stored under pressure in four 50-ton horizontal cylinders of the type known as bullets. On 13 July 1973 one of these bullets exploded whilst being filled from a rail car. An estimated 30 tons of liquid ammonia escaped from the vessel and a further 8 tons was lost from the rail car before the feed pump could be shut down. The air was still at the time of the spill and a large gas cloud formed, but within a few minutes a slight breeze arose, driving the gas over the perimeter fence and into a neighbouring township.

One man was killed instantly by the explosion. Others who were close to and in the direct line of the blast tried to escape but were overcome; two men managed to climb out of a storage tank and ran twenty-five yards before collapsing. The occupants of a control room about 80 yards from the explosion survived, including one man who was pulled inside with his clothes soaked in ammonia and covered with ice (ammonia chills sharply as it expands). They put wet cloths over their faces and after about half an hour were led to safety. Outside the factory, in the township, four people were killed immediately and two others died later. There were 18 deaths altogether, 65 people were treated in hospital and an unknown number by doctors outside the hospital.

At the time of the accident the tank car was filling No 3 and 4 tanks simultaneously. It was No 3 tank that failed, and fortunately an emergency valve in the line connecting the two tanks shut and held back the contents of No 4 tank. The temperature of the liquid ammonia was 15 °C and the pressure 90 psi. The vessels were designed to BS 1515 for an operating pressure of 250 psi. There was therefore no question of the vessels being subject to an overpressure.

The failure occurred when a disc-shaped piece of metal blew out of one of the heads. Figure 3.24 is an end-on sketch of the failed head showing the outline of the disc. The points marked A and B are, respectively, the sites of major and minor repair welds, whilst C was the origin of the brittle fracture. There was no obvious cause of fracture initiation. The day was sunny and the temperature 19 °C, and there would have been a modest thermal stress due to the ingress of relatively cold liquid. The fracture itself was undoubtedly brittle; there was no measurable thinning of the plate and the fracture surfaces showed the typical chevron marks. The blank for the head had been made up by welding together two carbon steel plates 23 mm thick, one large and one small. Subsequent tests showed that the large plate had an impact transition temperature of 115 °C, and the

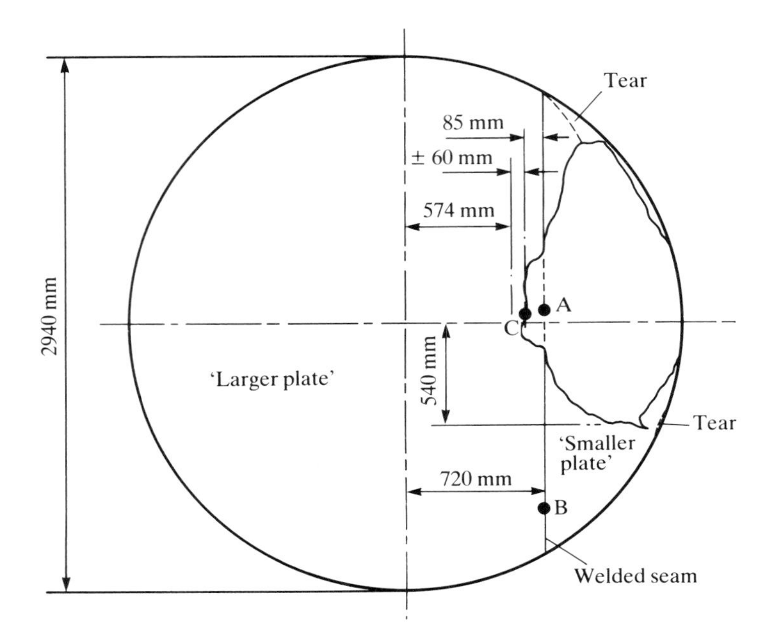

3.24 Brittle failure of the dished end of an ammonia storage tank.

small one 20 °C longitudinal and 35 °C transverse. The large plate had a hardness of about 200 Vickers and a small or zero elongation in the tensile test. Clearly this plate (which is the one in which the fracture initiated) was very brittle indeed.

The cause of this extreme brittleness is not at all clear. The head had been made by cold-forming the major radius and hot-forming the knuckle at about 800 °C. It had not been heat-treated subsequently and the vessel itself was not given a post-welding heat treatment because this was not required by the code. The steel had been ordered to a British Standard and the chemical composition conformed to requirements and did not show any peculiarities. Metallurgists investigating the problem suggested that the embrittlement could have been due to strain-ageing following the cold forming operation. However, many pressure vessel heads are made from similar steels in this way without disastrous consequences.

The reason for the repair welding is interesting. The bullet was subject to a mandatory periodic inspection, and the previous one had been carried out in late 1971. Normally a hydrostatic test was required, but the company obtained an exemption subject to an ultrasonic scan of the plate. The ultrasonic scan showed lamination in the larger plate of the dished head. Two areas of weld were scheduled for repair, either because of weld defects or because there were terminations close to the weld boundary.

One repair (the one marked B in Fig. 3.24) was made satisfactorily, but the other (A) gave trouble and eventually there was a substantial repair area 8 in long. Finally, a hydrostatic test was carried out, but stress relieving was not.

The residual stress field associated with the large repair weld was undoubtedly a major contributor to the disaster. The point of initiation was close to the repair, and the brittle fracture formed a closed loop, which is a rare feature and suggests that residual stress was the main driving force for the crack. Internal pressure then projected the broken-out piece for a distance of about forty yards; it finally collided with and ruptured an acid tank.

The owners of the plant (AE & CI; in earlier times, African Explosives and Chemical Industry) resolved that in future, formed heads, completed pressure vessels and repair welds should all be given a stress-relieving heat treatment. They also decided to provide gas-proof rooms in hazardous areas. This proposal is along the same lines as that of Lord Cullen following the Piper Alpha catastrophe, namely the provision of a temporary safe haven.

Generalities: catastrophes due to brittle failure

The two cases of catastrophic brittle failure described share two common features with the Liberty ships: the steel had a low notch-ductility at the operating temperature and serious defects were present which predisposed structural failure. In the case of the Liberty ships there were stress concentrations at the hatch corners, weld defects and fatigue cracks. The tiebars on the jacks of the *Sea Gem* had a sharp radius at the spade end together with surface weld runs that contained defects, combined with cracks and a hard heat-affected zone. The ammonia tank failure resulted from a combination of very brittle steel and a repair weld, giving a high residual stress field. In other words, the steel had been grossly misused in one way or another. Cases of pressure vessels that suffered brittle failure under hydrotest fall in a similar category: in almost all cases there was a combination of cracking with some degree of embrittlement.

There is an obverse side to this particular coin. The possibility of embrittlement due to post-weld heat treatment in a furnace has been mentioned earlier. During the investigation that followed the vapour cloud explosion at an ethylene plant, described in Chapter 1, samples were taken from two of the distillation towers that operated at subzero temperature. These were made of either 3½% or 5% nickel alloy steel, and impact tests showed that their notch-ductility was very low indeed. The subsequent history of these vessels is not known, but they had operated perfectly satisfactorily up to the time of the explosion. The ammonia tank at Potchefstroom had operated for four years with a dished end containing steel that had a transition temperature of 115 °C and a further two years with a severe, unheat-treated repair weld in the same dished end. Such experiences suggest that some structures can be very tolerant of embrittlement. In pressure vessels and piping this may well be due in part to the conditioning that the material receives during the hydrostatic test. The overpressure puts areas of stress concentration into a state of compressive stress, and this provides a guard against subsequent failure at the normal working pressure. Proof loading of machinery and other equipment performs the same function.

Problems arise when some unexpected mode of deterioration or failure occurs, like the lamellar tearing that led to the loss of the *Alexander L Kielland* platform. Experience brings such problems to light and makes possible specific provisions against them. At the same time, the general improvement in the cleanliness and notch-ductility of steel that has taken place during recent years means that the probability of catastrophic brittle fracture is progressively being reduced.

There is one other matter concerning maritime disasters that deserves comment. It will be recalled that after the *Sea Gem* fell into the water, a number of the crew went up to the helideck and refused to come down and board the liferafts, even though urged to do so. In other disasters men behaved in a similar way. On the *Alexander L Kielland* they congregated on the highest point of the wreck and would not board the lifeboats. Launching these lifeboats was indeed a hazardous operation, but the alternative was worse. In the case of the Piper Alpha, men gathered in the galley, again one of the highest points, and stayed there until the accommodation block fell into the sea. Some passengers on the *Titanic* were reluctant to leave the apparent safety of their quarters. A sizeable proportion of deaths was due to such conduct. Thirteen men from the *Sea Gem* lost their lives, and ten of these were those who stayed on board the wreck. Eighty bodies were recovered from the galley of the Piper Alpha after the tragedy.

This human trait needs to be borne in mind when seeking to minimise the loss of life due to disasters at sea. It could be argued that the problem is not that of providing a temporary safe haven, which is a straightforward design job, but how to persuade all the crew, not just the bold ones, to face the hazards and discomforts of the open sea.

Fatigue cracking

As well as being responsible for some major catastrophes – the Comet aircraft for example – fatigue failures cause a host of lesser breakdowns of plant and equipment, and are responsible for a high proportion of severe failures. It is, therefore, worth looking at the phenomenon in more detail.

Fatigue failure is an old enemy. It has been with us since the early days of the industrial revolution, when rotating machinery was first used on a large scale. A shaft carrying an overhung pulley could suffer many millions of stress reversals during its life, and the earlier type of fatigue test, the Wohler test, employed a loaded rotating cylindrical specimen. From such tests it was determined that in the case of steel there was a fatigue limit, that is to say, a stress below which no fatigue cracking would occur, and that this limit was in the region of one half the ultimate tensile strength of the steel. Pulleys, gears and the like were (and usually still are) attached to shafts using drilled holes or keyways, and such discontinuities were found to reduce the fatigue strength. From the amount by which the fatigue limit was reduced, the stress concentration factor associated with the discontinuity could be calculated. Thus, in earlier times, it was possible to design machinery and equipment with some confidence by taking stress concentration into account and providing a safety factor on the ultimate strength. The advent of fusion-welded structures has radically changed this situation by introducing new uncertainties about fatigue behaviour. To explain how this came about it is necessary to consider the mechanics of fatigue failure.

A notable feature of fatigue cracks is that they occur at a stress below the yield strength and show no evidence of plastic deformation. The cracks propagate along relatively straight lines at right angles to the direction of the principal tensile alternating stress. On a macro scale the fracture surfaces show coarse striations, but when examined at high magnification it is often possible to find fine regular parallel ridges which are taken to represent the extension of the crack during the individual stress cycle (Fig. 3.25).

There is no generally agreed model for the fatigue cracking process but the presence of microstriations does indicate that it must occur progressively by steps. Figure 3.26 indicates how this may happen. The stress cycle is assumed to be from zero to a maximum tensile value. As the stress approaches its peak, a cavity forms in that part of the plastic zone in which the hydrostatic tension is greatest, immediately ahead of the crack tip. At maximum load the ligament between the crack tip and the cavity fails, the crack advances by one striation, and the plastic zone reforms one striation forward.

According to fracture mechanics theory, the size of the plastic zone should be proportional to the crack length, that is to say, proportional to the square of the stress intensity factor K. So the crack growth per stress

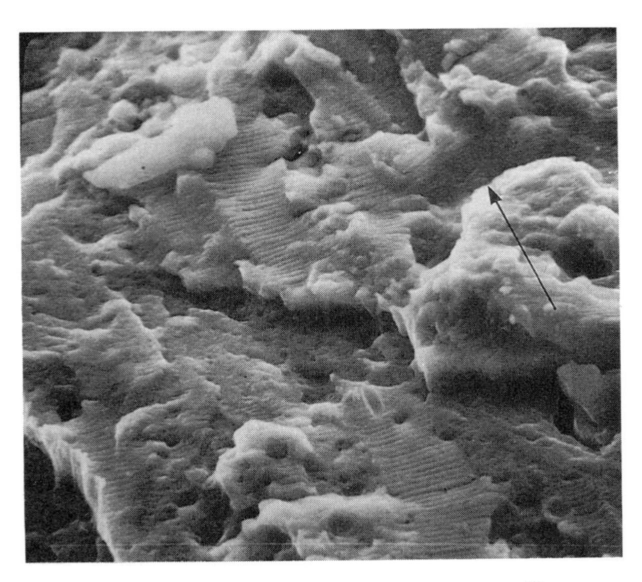

3.25 Striations on the surface of a fatigue fracture.[10] The arrow indicates the direction of propagation (× 1350).

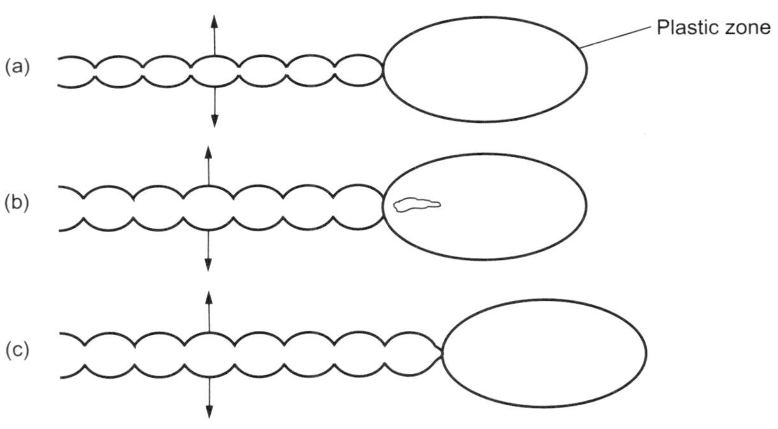

3.26 Fatigue crack growth; (a) start of half-cycle; (b) incipient failure; (c) end of half-cycle.

cycle should also be proportional to K^2, provided that the size of the cavity formed ahead of the crack is in proportion to the plastic zone size. Suppose that this proposition is correct. Then the crack growth rate is

$$\mathrm{d}a/\mathrm{d}N = c\,(\Delta K)^2 \qquad [3.6]$$

where c is a constant and

$$K = k(\pi a)^{1/2}\,\Delta\sigma \qquad [3.7]$$

$\Delta\sigma$ being the stress range and k being another constant. These equations may be integrated to give

$$\sigma^2{}_f\, N_f = \text{const} \times \ln(a_f/a_0) \qquad [3.8]$$

Here a_0 is the initial crack length and σ_f, N_f and a_f are the stresses, number of cycles and crack length at failure. For test specimens having a constant width and constant initial crack size the right hand side of this equation is constant, and a plot of log (failure stress) against log (cycles to failure) should give a straight line with a slope of $-\frac{1}{2}$.

Fatigue curves for three types of carbon steel specimen, plain, with a drilled hole and with a fillet-welded attachment are plotted in Fig. 3.27. The log–log plot is indeed a straight line, but in no case is the slope of the curve equal to $-\frac{1}{2}$. In the case of samples with a fillet weld, then there is likely to be a crack-line defect at the toe of the weld, as illustrated in Fig. 3.28. With such a configuration the fatigue life is occupied almost entirely by crack propagation, and the lower slope (about $-\frac{1}{3}$) simply reflects the over-simplified nature of the model proposed here.

In the other two cases it would have been necessary to develop an initial crack. When a metal is subject to alternating stress at levels below the yield

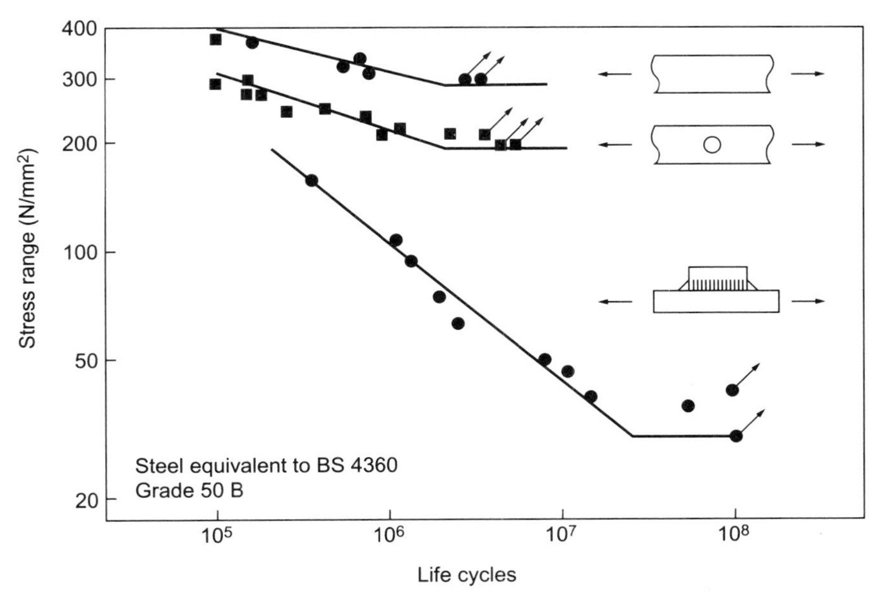

3.27 Stress range versus cycles to failure in a fatigue test for mild plate, plate with a drilled hole and plate with a transverse fillet weld.

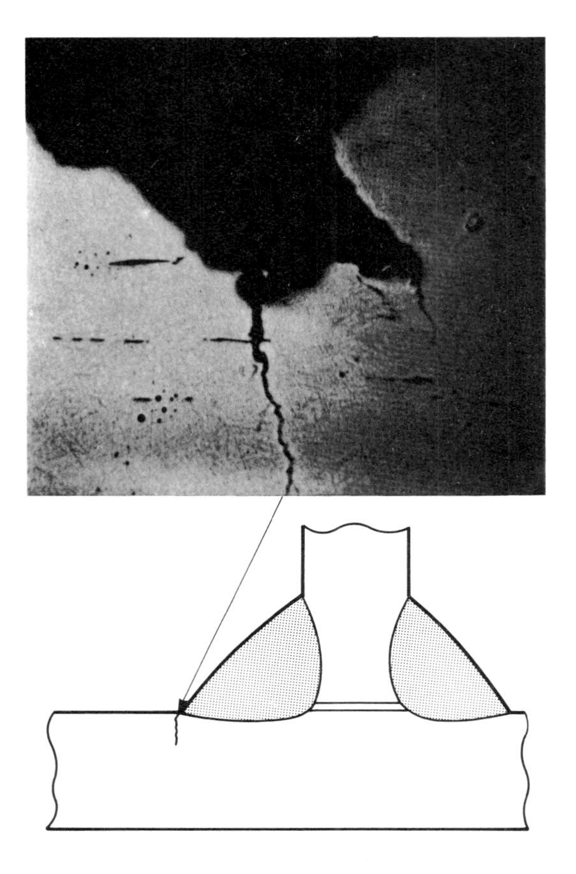

3.28 Fatigue cracking initiated at a slag-filled discontinuity at the toe of a fillet weld.

stress, slip occurs across shear planes, and after a large number of cycles this generates steps and eventually cracks at the surface. In such circumstances much of the fatigue life is spent in starting the crack. The lower the stress, the greater the number of cycles required to form a crack, so the fatigue life at low stress is increased to a greater degree. Hence the stress versus the cycles-to-failure line becomes less steep. Plots for welded joints other than transverse fillet welds lie between the two extremes, and standard design stress figures are available based on the experimental data.

Improving the fatigue resistance

One factor that affects the fatigue performance of welded joints is the presence of residual tensile stress. The effect of this locked-in stress is that when the member containing the joint is subject to an alternating stress that is partly compressive, the regions local to the weld suffer a completely tensile stress range, with a correspondingly greater risk of damage. As would be expected, stress-relief is beneficial in those cases where 50% or more of the stress cycle is compressive. Design rules for structures (e.g. BS 5400) do not give any bonus for stress-relieving, but such heat treatment is used to some extent in vehicle construction.

A more general benefit is obtained by cold-working the surface around the toe of the weld in order to induce a localised compressive stress. Alternatively the toe of the weld may be ground to reduce the stress concentration effect and to remove any crack-like defects. The effect of these two alternatives on the stress range versus cycles to failure curve is illustrated in Fig. 3.29. The ratio *R* indicated on the figure is that between the minimum and maximum stresses, and since tensile stresses are conventionally assigned a positive sign, this means that the stress range was entirely tensile.

Cold-working for the test here recorded was carried out by hammer-peening, using a pneumatic tool. Taking the stress for failure at two million cycles as the measure of fatigue resistance, which is a commonly used criterion, peening more than doubles the permissible stress range.

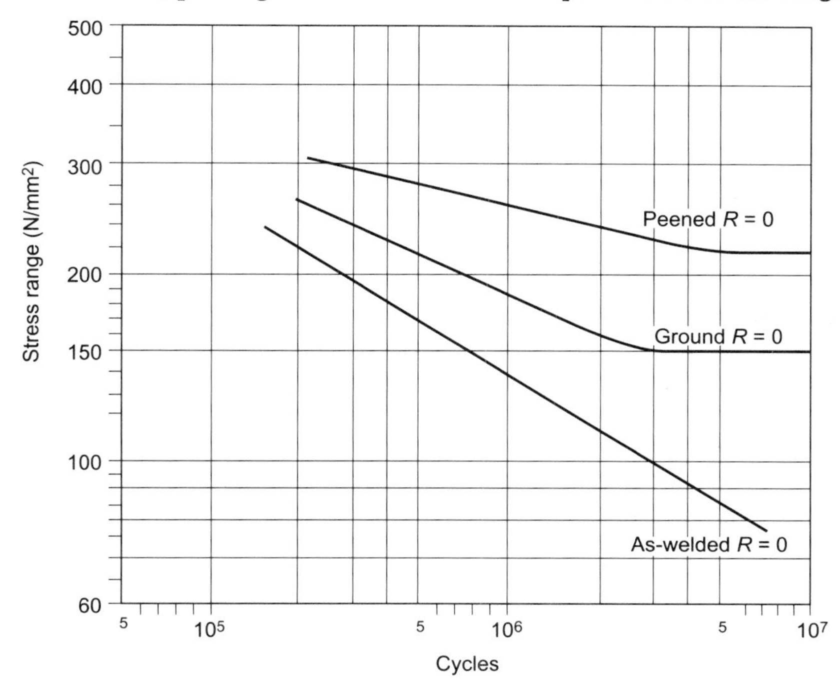

3.29 A comparison of the fatigue strength of fillet welds in the peened, ground and as-welded condition.

Although it is technically desirable, this technique has been little used in practice. One reason is the noise that it creates. On a recent job, the operator was (even with ear muffs) allowed to work for only 15 minutes in any one hour. The other matter for concern is the effect of cold-working on the notch-toughness of the steel. This is of little consequence for thin sections, but in the case of thick sections used for North Sea structures, users have been reluctant to accept the risk. In fact, only a thin surface layer is affected and it is questionable whether there would be any real increase in the susceptibility to brittle fracture.

Recollections of the *Sea Gem* disaster still cast their shadows, however, and the more expensive and less effective technique of grinding has been specified for North Sea rigs in critical areas such as the node connections. Grinding is done with a burr tool or a disc grinder – normally the disc grinder. Specifications typically require a groove of minimum depth 0.8 mm. Grinding has the advantage of being easier to apply and very much easier to inspect than peening. There are other methods of improving fatigue resistance, including remelting the toe area by means of a gas tungsten arc torch, but these have been little used in practice.

Service failures

Not all fatigue failures start at welds. A refinery manager well known for his dislike and distrust of contractors, complained that his recently built hydrodesulphuriser was very noisy, and at first this was thought to be a typical grouse. One day, however, an operator noticed a fine crack running around the inlet flange to the reactor. The unit was shut down and the flange removed; when examined it was found to a millimetre or so from complete rupture. Further investigation showed that the refinery manager had in fact been right. The feed/effluent heat exchanger for the reactor had been so designed as to set up an organ-pipe or standing wave oscillation of fluid in the tubes, and the resulting vibration had been transmitted down the transfer line to the flange, which failed at a sharp corner radius. Noise can indeed indicate an unsatisfactory state of affairs in process equipment.

Maddox[10] gives many case histories. Figure 3.30 shows two types of fatigue failure of a chord-to-brace joint in a fixed offshore structure. The brace in question was located near the surface and was therefore subject to substantial fluctuating loads due to wave motion. Horizontal movement of the chord gave rise to axial loads on the brace, as indicated in the sketch. This in turn initiated cracks starting from the toe of the weld on the chord side (photo (a)). Vertical movements on the other hand resulted in displacement of the brace, and initiated a fracture through the brace (photo (b)). This problem was largely overcome by locating the braces at a

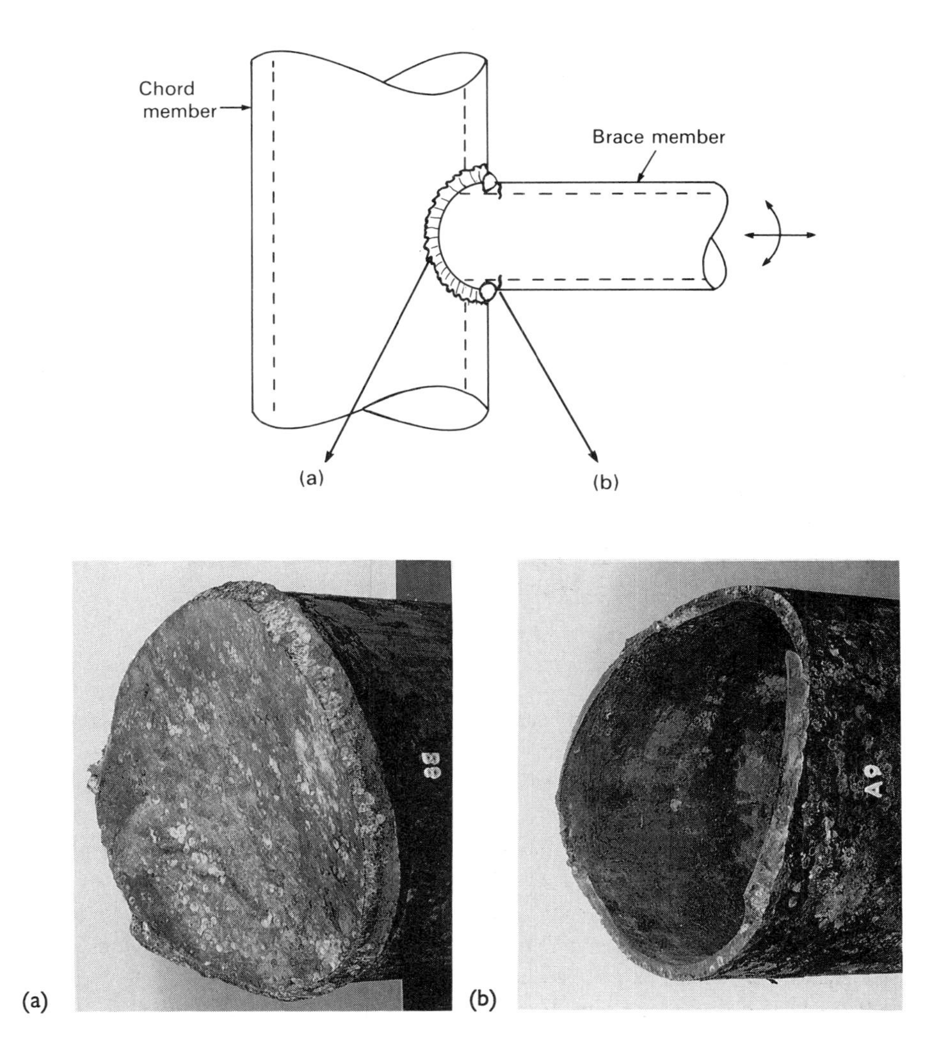

3.30 Fatigue failures of chord-to-brace welds in offshore structures: (a) from weld toe into chord; (b) from weld toe into brace.[10]

lower level, but a few similar fatigue cracks have occurred in other areas, due no doubt to the considerable height of waves in the North Sea.

Maddox also draws attention to the hazards due to fabrication aids. One well-known example is the cleats that are welded to the outer skin of pressure vessels in order to align strakes or longitudinal welds. Most specifications require that such attachments be cut off, the surface ground smooth and tested for cracks, but whereas great care is taken in making procedure tests for the main seam welds, there is virtually no control over temporary attachment welds, which can provide equal and sometimes greater hazards.

Figure 3.31 illustrates two service failures that resulted from fabrication aids. The upper figure is a sketch of a section of a longitudinally welded tubular member which had been subject to alternating stress. The weld had been made on to a backing bar, whilst the backing bar itself was made up of short lengths of strip welded end to end, the pieces being joined by partial penetration welds. Fatigue cracks initiated at the unfused discontinuity in the backing bar and propagated through the pipe, as shown.

The other diagram shows a node area in an offshore structure. The braces are joined to the vertical member by butt welds, and to make

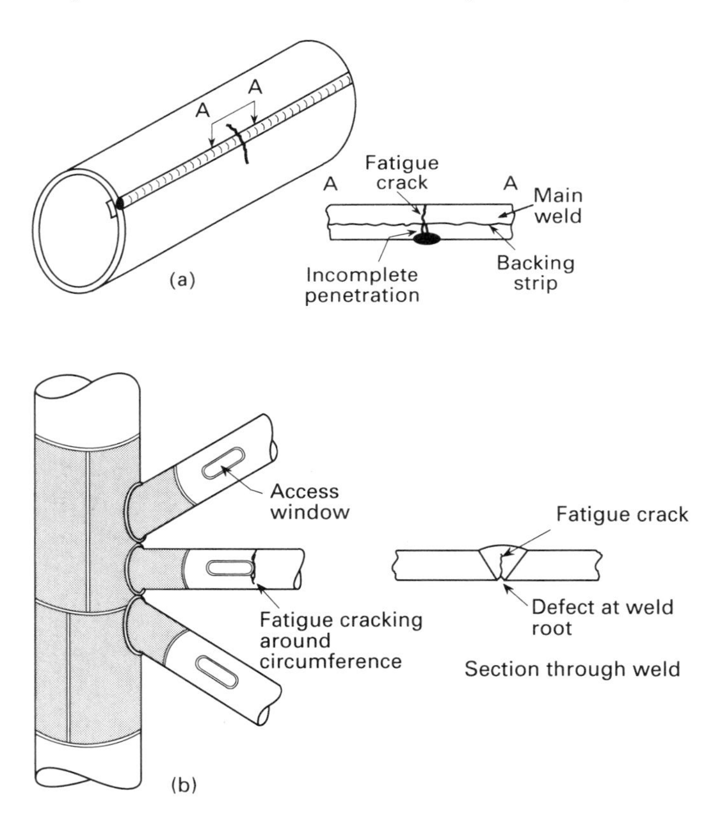

3.31 Fatigue cracking in service: (a) propagating from partial penetration weld in backing strip; (b) initiated by defect in most of single-side weld made to close up access windows in brace of offshore structures.[10]

inspection of the root of these welds easier, a window was cut into each of the braces. Finally, the holes were repaired by welding in a closure piece. However, owing to the poor fit and an awkward welding position, the welds contained root defects. These propagated to the surface, and although the cracks were found before any catastrophic failure occurred, repairs were difficult and costly.

Repairing the damage

There is of course a degree of unpredictability about the fatigue behaviour of welded joints, and any welded structure subject to fluctuating loads must be regularly inspected for cracking. The traditional means of dealing with a crack was to drill a hole at the tip. This method was used by de Havilland during the construction of the early Comet aircraft, and although it had no effect on the ultimate failures, the practice was condemned by those enquiring into the disasters and was abandoned by the manufacturers. The principle adopted by de Havilland, and that which now governs generally in engineering construction and maintenance, is to replace a cracked part with a sound one which is strong enough to resist cracking in the future.

Fatigue cracking has been a serious problem with the steel decks of some post-1945 suspension bridges of the box-girder type. The cracks which afflicted the Severn bridge have been well documented, as have the methods of repair.[11] In the present context these are of interest as a means of preventing a serious extension of fatigue cracking and possible fracture.

The bridge across the river Severn forms part of the motorway link between London and South Wales. The structure consists of a series of welded steel boxes which are bolted together and suspended by cables from two towers. The boxes support a steel deck which is stiffened by longitudinal trough-shaped members. It is the deck, which is subject to fatigue stress owing to vehicles crossing the bridge, that has suffered fatigue cracking. Three types of weld were affected; fillet welds joining a temporary attachment to the base of the stiffener troughs; welds between stiffeners and crossbeams; and welds between stiffeners and the underside of the deck plates. It was required that the repairs should also strengthen the joints to give a service life of 120 years, that they should be carried out with minimum effect on the traffic flow, and without adding significantly to the suspended weight. Access was through 600 × 300 mm hatchways. These were severe conditions, and the weight restriction in particular meant that a simple solution such as adding extra stiffeners was out of the question.

Repair methods

The first category of failure is sketched in Fig. 3.32. This is another failure caused by a construction aid. Sections of the suspended structure were fabricated upstream on the river Wye and then floated down to the required position. To do this it was necessary to install a temporary diaphragm, known as the flotation diaphragm, to one end of the section. As shown in the figure this diaphragm was welded to the bottom of the stiffener troughs, and it was not removed after erection. In consequence substantial bending loads were applied to the welds, and cracks appeared within five years of the bridge opening. As an interim step, the diaphragms

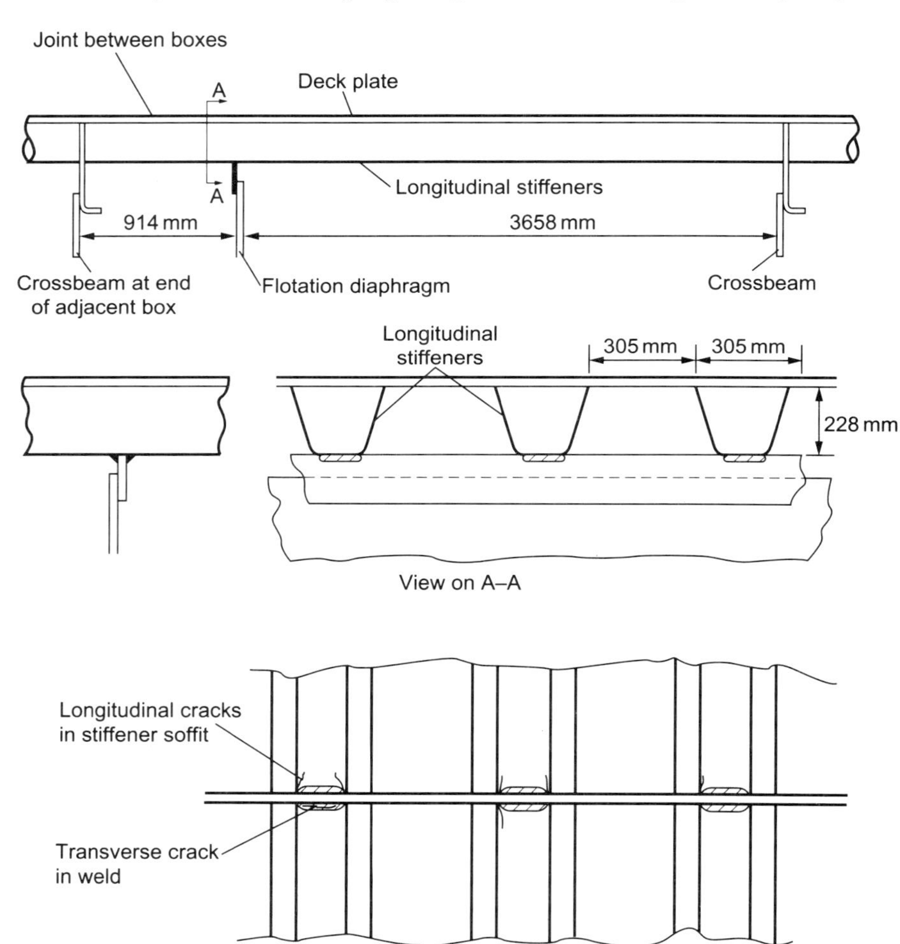

3.32 Fatigue cracking associated with weld between flotation diaphragm and stiffener for deck of Severn bridge.[11]

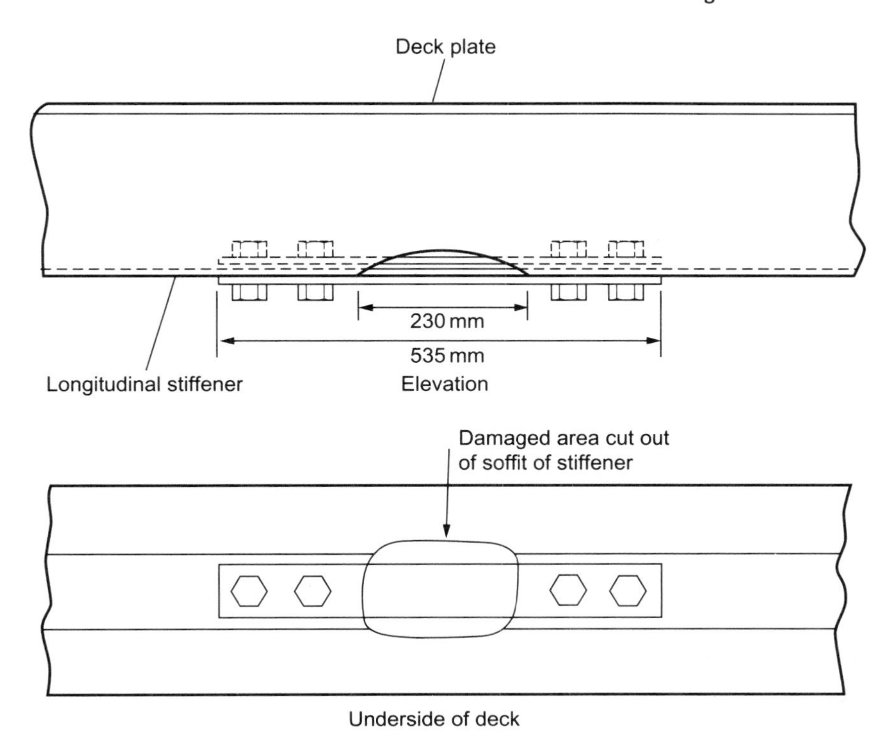

3.33 Repair of flotation diaphragm welds.[11]

were cut below the welds and the cracks were welded up. It was calculated that such repairs would have a life of three to five years.

The permanent solution was to cut out damaged areas and to bridge the hole with a plate which was attached by high tensile friction bolts, as shown in Fig. 3.33. Fatigue tests were carried out on laboratory specimens and on a full size deck panel; these indicated a life in excess of 120 years. A large number of these bridging plates have been installed and are performing as planned.

At a later stage, somewhat less than 20 years after construction, cracks appeared in the fillet welds connecting the longitudinal stiffeners with transverse crossbeams. The layout of these members is shown in Fig. 3.34. Similar cracks were found on two other crossings forming part of the same motorway system, the Wye Bridge and the Beachley Viaduct. Static tests using strain gauges in the positions indicated in the figure showed a wide range of results, varying over a factor of three. Fatigue tests also showed a considerable scatter. Such variation was thought to be due to differences in the geometry and fit-up of individual joints. In devising a repair and strengthening method, it was clearly necessary to minimise variations in

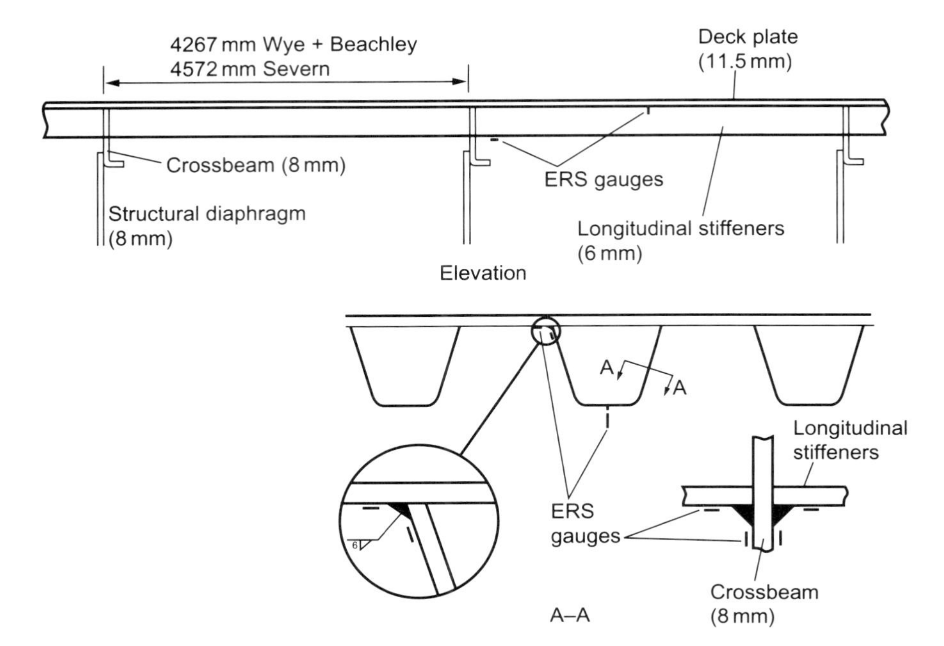

3.34 Welds between stiffener and crossbeam and between stiffener and deck of Severn bridge. Both suffered fatigue cracking. The figure shows the location of electric resistance strain (ERS) gauges.[11]

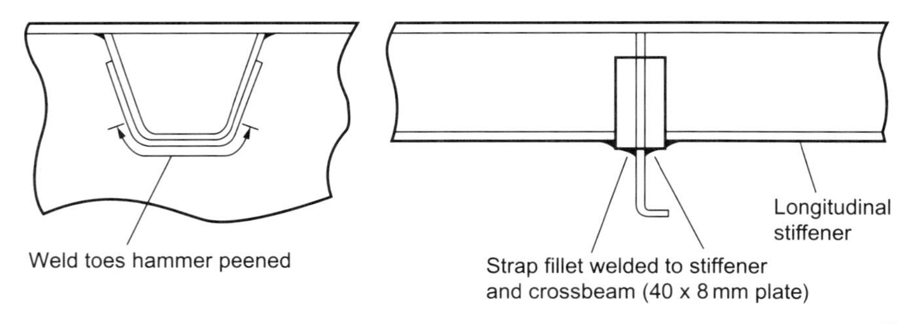

3.35 Repair and strengthening of crossbeam to stiffener joint, Severn crossing.[11]

fit-up. The scheme finally adopted was to heat 40 × 8 mm strips of steel and then wrap them round the stiffeners on either side of the crossbeams, thus ensuring a good fit. The strips were fillet-welded to stiffener and crossbeam, and the weld toes were hammer peened as indicated in Fig. 3.35.

Cracking was also found in the fillet welds connecting stiffeners to the underside of the deck plate. Laboratory tests showed that the required life could be obtained by replacing the fillet by a multipass penetration weld

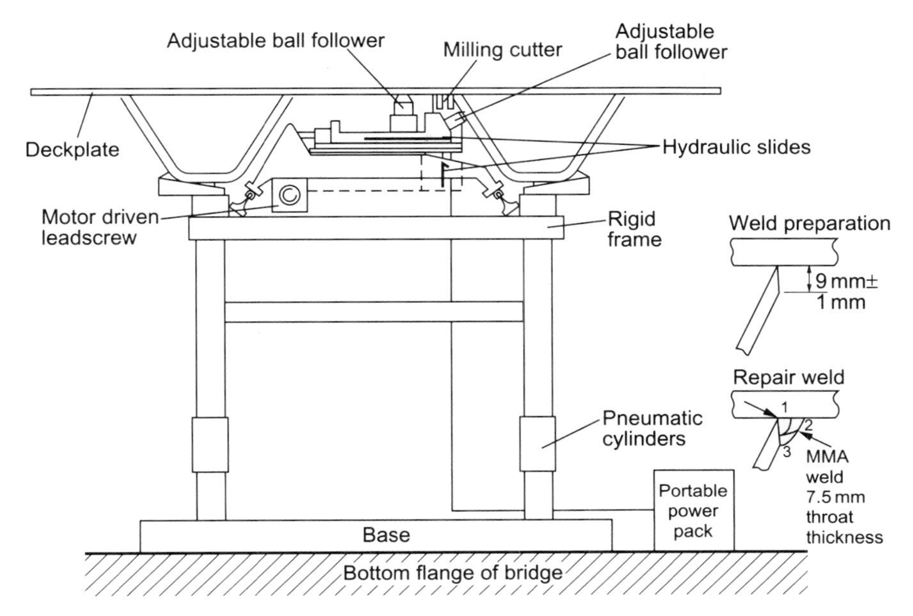

3.36 Machining device for making the edge preparation of the stiffener to deck weld repairs, Severn crossing.[11]

with a 7.5 mm throat thickness. However, a machined rather than a ground edge preparation was required to obtain this result. The problem was to devise a machine that would accommodate not only the static variation in profile along a stiffener, but also the movement caused by traffic flow. Eventually a device mounted on pneumatic cylinders was developed (Fig. 3.36). This was equipped with followers that maintained the machining tool in a constant orientation relative to the joint. After machining, the weld was made manually with coated electrodes.

All three of the repair methods described above have proved to be practicable, and are used for *ad hoc* repairs and as part of a general programme of strengthening the decks of the three crossings concerned.

Similar types of cracks have been found in the steel decks of other bridges. In no case however have they constituted a threat to the structural integrity of the bridge; rather they have given rise to a maintenance requirement. The fatigue design curve for welded bridges in BS 5400 for example, is based on tests showing that a high proportion (better than 97% in the case of BS 5400) of welds will survive the design life without cracking. This implies, however, that a small percentage *will* crack, so that the maintenance problem is, in effect, part of the design.

Cracks appeared in the steel decks of the Severn crossing earlier and more extensively than expected. This was because they were part of a pioneering project. Much has been learnt about the problem in recent years, such that there is every reason for confidence in this type of construction.

Explosions

The catastrophe of war is fuelled by explosions, but in this chapter we are concerned only with those that occur in civil life. They include the explosions of boilers, which were discussed in Chapter 1. It is obviously desirable to know something about these phenomena, even though this will not necessarily point to any means of preventing them.

The behaviour of hydrocarbon–air mixtures

A high proportion of the domestic comforts of contemporary civilisation depend directly or indirectly on the controlled combustion of hydrocarbons. In the majority of countries (France is exceptional in this respect) most electricity is generated in power stations that are fired by coal, heavy oil or natural gas. Central heating boilers usually employ oil or gas as an energy source. The family motorcar depends on controlled explosions of hydrocarbon–air mixtures for motive power. And for the present, at least, the rate of discovery of crude oil keeps pace with rising demand, while the known reserves of natural gas increase in quantity more rapidly than does consumption. It is in the refining of these essential sources of heat and power, and in transporting them from place to place, that the problems arise.

The Bunsen burner (Fig. 3.37) provides a good example of steady controlled combustion. A jet of methane gas comes out of a small hole opposite the air inlet, and the two gases mix. The mixture then passes up the vertical tube, where any turbulence is smoothed out. At the outlet of the tube the gas mixture burns radially inwards, and, combined with the upward flow, this results in a conical flame front, inside which is the familiar blue cone of unburnt gas, and outside the feathery plume of hot burnt gas. All is in balance; the rate of advance of the flame front equals the rate of upward flow. However, if this upward flow rate is reduced too much, an instability may arise and the flame burns back down the tube and ignites the methane gas as it comes out of the small orifice. Any further burn-back is unlikely, however, because the flame would be quenched by the cylindrical wall of the orifice.

To ignite a hydrocarbon–air mixture two preconditions must be met. Firstly, the mixture must be raised to its auto-ignition temperature. This is the temperature at which the combustion reaction is self-sustaining.

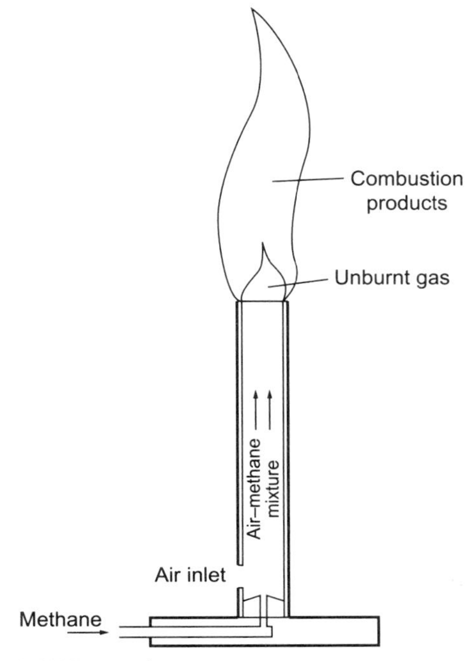

3.37 Bunsen burner.

Secondly, the heated volume must be large enough such that the rate of heat generated by combustion is greater than the rate of heat loss to the surroundings. In the case of volatile or gaseous hydrocarbons these conditions are easily met, and a match or small electrical spark is sufficient.

Deflagration and detonation

The conditions described above represent the steady combustion of a laminar gas mixture flow. To achieve this condition it is necessary to provide a controlled rate of flow of the mixture. When a cloud of gas mixture is ignited there is no such control, and the flame front advances in a disorganised manner. Two extreme conditions may be conceived. The first, which may occur for a short period after ignition, is a steady, relatively slow advance of the flame front. In the second extreme, the rate has become supersonic. The first condition is known as deflagration, and the second detonation.

Detonation

This type of explosion occurs when a shock wave travels through a suitable material. A shock wave is a pressure wave travelling at a speed greater than that of sound. The material ahead of the wave is substantially undisturbed, and at the wavefront itself there is an abrupt rise of pressure. This pressure rise heats the material to its auto-ignition temperature or beyond, making the condition self-sustaining.

High explosives such as TNT behave in this way. A high explosive bomb consists of the main charge enclosed in a steel case, usually cylindrical. A tube or pocket running through the TNT (or other high explosive) contains a less stable chemical compound, the gain, and embedded in the gain is a substance that will detonate when struck or suddenly heated. The detonator creates an initial shock wave that is amplified by the gain sufficiently to detonate the main charge. At the end of this process the solid explosive has been converted into a mixture of gases at exceedingly high pressure and temperature. A gaseous mixture of hydrocarbon and air may be detonated in the same manner. Calculated initial temperatures and pressures resulting from such an explosion are about 3000 °C and 18 atmospheres. Expansion of this hot compressed gas mixture takes place with high velocity and is preceded by a shock wave. Eventually the pressure behind the shock wave decays to atmospheric pressure and the shock wave becomes a sound wave. In the meantime, obstacles in front of the shock or blast wave will have been destroyed or damaged.

Early in 1941 the possibility of developing an atom bomb was being discussed in Britain and the USA. One problem was that although such a bomb would generate a large amount of energy it would not produce the gases which result from the explosion of, say, TNT. Accordingly, Sir Geoffrey Taylor, an expert in fluid dynamics, was asked what mechanical effects would be produced by a sudden release of energy. At the time, his response was a military secret, but in 1950 the paper was declassified and published in its original form.[12] The prediction was, of course, that the mechanical effect would be the same as that due to a chemical explosion: a volume of air would be heated to produce very high temperatures and pressures, and expansion of this volume would be preceded by a shock wave. Radiation effects were not considered in this study.

Assuming that the energy of the explosion E was generated at a point, the gas cloud would expand in spherical form. In this case the maximum pressure p_{m} behind the shock wave was calculated to be

$$p_{\mathrm{m}} = 0.155\ E/R^3 \qquad [3.9]$$

where R is the distance from the source of the explosion. When the first atom bomb was exploded, it was found that this simple expression predicted the results remarkably well. Comparisons with measurements of blast pressures produced by chemical explosives showed that the assumption of a point source caused the theoretical pressures close to the explosion to be too high, while at large distances the predictions were too low. In the intermediate range the trend was similar but the pressures calculated for the nuclear explosion were about half the actual pressures generated by a chemical explosion of the same energy. This is consistent with the idea that the gas produced by chemical explosions makes a positive contribution to blast pressure.

Figure 3.38 shows such a comparison, but in this case the multiplying factor in the equation has been doubled to give

$$p_m = 0.3\ E/R^3 \qquad [3.10]$$

The measured pressures were made for a mixture of explosives that gave an energy release of 5024 J/g. The charge weight was 206 kg, giving a value for E of 1.202×10^9 J. Expressing the pressure p_m in atmospheres (1.013×10^5 N/m^2) and radius R in metres, this leads to

$$p_m R^3 = 3607.3\ \text{J} \qquad [3.11]$$

At distances between 3 and 8 m the agreement between Taylor's equation (multiplied by 2) and measured pressure is very good.

The quantity $(E/p_0)^{1/3}$ where p_0 is atmospheric pressure, is called the 'characteristic length' R_0 of the explosion. The 'characteristic time' of the explosion is R_0/C_0, where C_0 is the velocity of sound at room temperature and atmospheric pressure. The characteristic length is an approximate relative measure of the radius from the point of explosion within which damage would be expected. For example, in the case of the mixture of explosives referred to above.

$$R_0 = \left(\frac{1.202 \times 10^9}{1.013 \times 10^5}\right)^{1/3} = 22.8\,\text{m} \qquad [3.12]$$

Extrapolation of a log–log plot of the measured values indicates that the blast pressure would fall to one atmosphere at about 48 m distance. However, calculations of the extent of damage due to explosions are notoriously unreliable, and the best guide is obtained from empirical observations.

The detonation of hydrocarbon–air mixtures[13,14]

One method of detonating gas–air mixtures is to enclose the gas in a plastic balloon and set it off with a charge of the explosive tetryl. Usually the volumetric ratio of hydrocarbon to air is such that enough air is present to oxidise the hydrocarbon completely; this is known as the stoichiometric

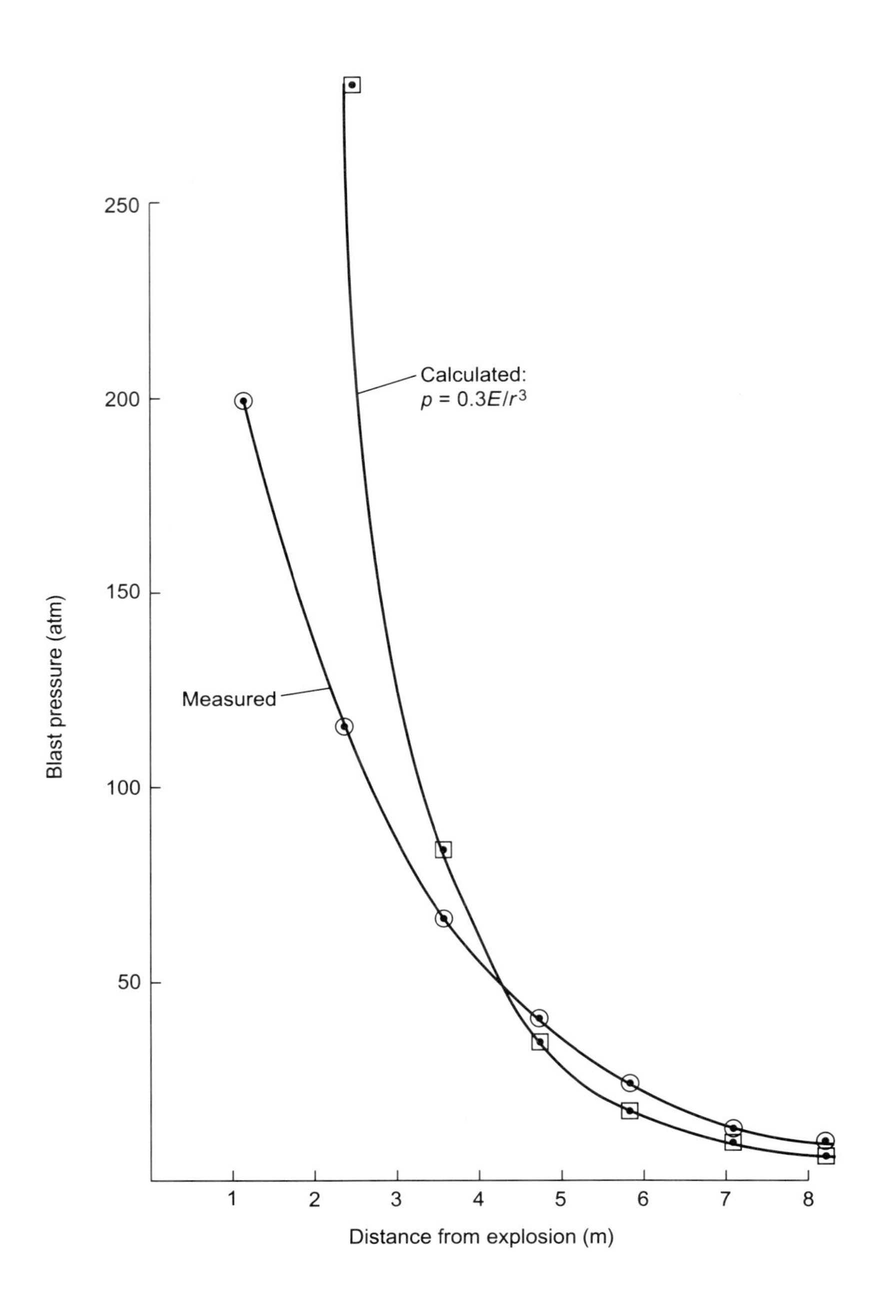

3.38 Blast pressure versus distance from explosive. One standard atmosphere is equal to 1.01325×10^5 N/m^2.

ratio. The size of the charge required to cause detonation is a measure of the detonatability of the mixture. Much of the experimental work in this field has been devoted to methane–air mixtures, since methane is used and handled on a large scale. Figure 3.39 shows the results obtained by one set of investigators.[13] In this instance the figure for methane was obtained by diluting oxygen–methane mixtures with successively larger amounts of nitrogen, and extrapolating the results to that corresponding to the composition of air. This gave an amount of 22 kg of tetryl. Others initiated detonations in stoichiometric air–methane mixtures with between 1 and 4 kg of high explosive. Even with these figures, methane is much the most stable of the hydrocarbon gases. This is in accordance with experience; there do not appear to have been any vapour cloud explosions due to the release of methane. There have indeed been serious explosions at natural gas processing plants, but these could have been associated with the presence of propane. Confined explosions caused by methane leaks are, on the other hand, relatively common, and these will be mentioned later.

Relative detonatability is related to the chemical stability of the compound in question, and this in turn is dependent on the ratio of the number of hydrogen atoms in the molecule to the number of carbon atoms. Carbon is quadrivalent, so that in methane, CH_4, all valence bonds are satisfied and the gas is relatively stable. At the other extreme, acetylene, C_2H_2, has a hydrogen/carbon ratio of 1 and is notoriously

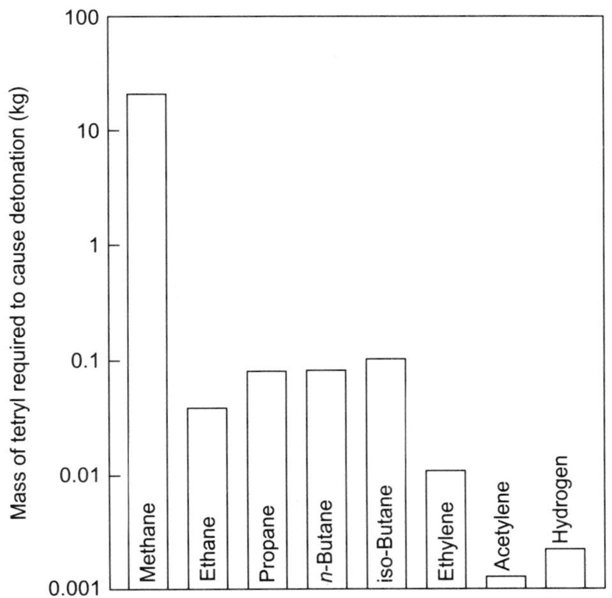

3.39 Detonatability of various hydrocarbon–air mixtures.[13]

unstable. Ethylene, C_2H_4, is also easily detonated, and there have been serious vapour cloud explosions due to ethylene emissions. Cyclohexane, which was responsible for the Flixborough explosion, has the formula C_6H_{12} and like ethylene, has a hydrogen–carbon ratio of 2.

Unconfined vapour cloud explosions

The most recent source of information on vapour cloud explosions is D G Mahoney's survey,[15] which was reviewed and analysed in Chapter 1. Other sources are listed in Baker and Tang's book on explosions.[13] Accidents in which there was a large-scale release of gas were examined by Gugan,[14] for the period 1921–77. It was found that 60% had caused blast damage over a wide area, but it was considered that this damage was due to deflagration rather than detonation. In some cases the vapour cloud failed to ignite, and in others the vapour burnt but did not have any blast effect.

It is not easy to account for the damage that does occur. In most cases the vapour cloud comes from the rupture of piping, less frequently from the explosion of a pressure vessel, and less frequently still because an operator left a valve open to atmosphere or made some similar error. Typically the cloud spreads until it reaches a source of ignition, which may be an open flame or an electric spark. The steady burning velocity of a hydrocarbon–air mixture is about 0.5 m/s, whereas that of a detonation is calculated to be about 1800 m/s. There is no detonator source in a normal process plant, so how is it possible for the flame front to accelerate from 0.5 to 1800 m/s? At present there is no convincing answer to this question. The pressure build-up due to combustion does not in itself provide a mechanism; it is quite possible to envisage a pressure wave of constant height in company with a flame front of constant velocity. Turbulence of the flame front due to solid boundaries has been proposed as a means of acceleration, but the vapour clouds with which we are primarily concerned often have no solid boundary. The type of damage produced by vapour cloud explosions is typical of that produced by a detonation blast wave. It must therefore be accepted that in some way, as yet to be adequately explained, ignition of hydrocarbon–air mixtures can result in an explosion.

Size may be a factor. It is perfectly safe to burn cordite in small quantities, but if a large amount is ignited it will explode. In gas–air mixtures, a finite time is required to develop the conditions for detonation, and the larger the vapour cloud the more likely it is that these conditions will obtain.

One of the necessary preconditions for an explosion is that there must be reasonably good mixing of hydrocarbon and air. In most cases the escape is violent and the jet of fluid will be turbulent, such that the air will be entrained. In many cases there is then a significant period before the mixture ignites, sufficient for some diffusion to take place. It is very unlikely, however, that a stoichiometric mixture will be achieved and there is likely to be an excess of hydrocarbon. Consequently the explosion energy is substantially less than that of the mass of hydrocarbon released. For actual vapour cloud explosions the proportion of material contributing to the explosion has been estimated as between 1 and 10% of the total. At Flixborough, it has been calculated that of the 30 tons of cyclohexane released, only 5% exploded. The unburnt hydrocarbon is dispersed, or in some cases is consumed in a slow burn.

Confined explosions

Here there is no difficulty in supplying a simple explanation of what happens. Owing to the ignition of a hydrocarbon–air mixture, or sudden evaporation of a liquid, or some other cause, there is a rapid rise of pressure inside the container. If vents are provided, then the peak pressure will depend on the balance between internal volume increase and rate of loss of gas through the vents. For example, if there is a gas leak in a house, and if the gas–air mixture is ignited, the internal pressure can build up to eight atmospheres, at which pressure the house will be demolished. Normally, however, windows will be blown out first and, by acting as a vent, this may protect the walls.

Vents in the form of rupture discs are often used to protect equipment from the effect of an internal overpressure or explosion. Rupture discs are of small diameter compared with the container and are made of relatively thin corrosion-resistant metal. If the internal pressure exceeds a specified limit, the rupture disc bursts rather than the vessel itself. These items do not always work as intended. The reactor used for producing batches of resin for industrial coatings in a Cincinnati works was fitted with such a disc. Operators were using solvent to clean the reactor before putting in a new charge. However, there was some residual heat in the vessel, solvent was vaporised and the pressure built up to a point where the disc burst, releasing a jet of vapour. Unfortunately the reactor was located in a building, and the vapour cloud accumulated in the roof until it hit an ignition source. The explosion wrecked the reactor building and damaged neighbouring buildings and stores at a cost of over US$23 million.

This particular episode was not unusual. The most common causes of internal explosion in process plant are excessive evaporation and processes that run out of control. For example, where a fluid catalytic cracker unit was being brought back on stream after a maintenance period, hot oil was introduced to a pressure vessel. However, the drain valve at the bottom of this vessel had not been properly closed and water had accumulated in it. When the hot oil hit the water there was an explosion and the vessel burst. The oil so released burst into flames and caused serious damage.

Internal explosions due to uncontrolled chemical reactions take place from time to time, although they cannot be rated as common. One such occurred in a plant making a flame retardant. This process unit, which is located in Charleston, South Carolina, was starting up after a maintenance period, and operations had reached the second step of a batch process. However, raw material used at this stage was contaminated with water that had been used during maintenance. At the same time the water flow to the reflux cooler in the reactor had been reduced because of blockage. Consequently there was a build-up of temperature, and rapid decomposition. The reactor burst, causing mechanical damage and spreading fire through the plant.

An internal explosion in a pipeline provided some useful lessons. This line was carrying fuel gas to an industrial area and it became necessary to increase the flow rate. So the pressure in the line was increased, the planned level being within the capability of the pipe, but above the original hydrotest pressure. As the pressure was being raised, there was an explosion and a large fire, which fortunately did not cause any serious damage, nor were there any casualties. An electric resistance welded pipe from a reputable manufacturer had been used for this line, and examination of the ruptured portion showed two significant features. Firstly, the pipe had been dented at some stage, probably by a mechanical digger during the pipe-laying operation. Secondly, the initial failure had been a split along the centre of the weld, and the adjacent length of pipe had split wide open, with 45% shear fractures. The most likely sequence of events was thought to be as follows:

1 The rise in pressure caused the dent in the pipe to pop out.
2 This event cracked the weld and gas started to escape, being ignited in the process.
3 Air was ingested into the pipe, causing an internal explosion.

Impact tests were carried out on the electric resistance weld, the V-groove being machined along the centre-line of the weld metal. The results were very low indeed and showed that the weld was almost completely brittle. There was no obvious reason why this should be so; the microstructure of samples taken from undamaged parts of the pipe showed no unusual features and were free of cracks and non-metallic inclusions. Samples of welds taken at random from electric resistance welded pipe of various sizes, however, also gave low values, not quite so low as the burst pipe but well below the normally accepted level. So the initial brittle fracture of the weld was not due to an extreme divergence from the norm, but rather to an unusual distortion resulting from the pressure rise acting on a damaged section of pipe.

The second point to note is that following the initial fracture, the escaping gas ignited. Such ignition sometimes occurs following the bursting of a pipe or pressure vessel, but the explanation is not straightforward except, of course, where the fluid is at a temperature above the level required for auto-ignition. Suppose that the work done in fracturing the metal is converted into heat, and there is no heat loss to the surroundings, then the temperature rise should be equal to the fracture stress divided by the specific heat per unit volume of the metal in question. Assuming the fracture stress to be the mean of yield stress and ultimate stress, then the temperature rise is calculated to be about 100 °C, far too low to ignite a hydrocarbon gas. On the other hand, if the fracture stress under shock loading conditions (which must have been the case here) were very much higher than the stresses measured at low strain rates in a tensile test machine, then the required temperature could well be reached. This notion is consistent with observations made earlier in this chapter about brittle fracture due to dynamic loading. It is also consistent with the fact that freshly broken impact test specimens are very hot indeed, as many young metallurgists have discovered to their cost.

Some ruptures are not followed by immediate ignition of the hydrocarbon. It is these that may cause vapour clouds to form and eventually explode. It is consistent with the argument set out here that ruptures associated with shock loading or (as in the case of the Piper Alpha accident) by impact will result in ignition, while other types of failure may not.

The third point of interest is that for an internal explosion in the pipe to occur, it was necessary for air to be drawn in. How this could have happened is a matter for speculation. It is possible that the sudden fracture set up pressure waves within the pipeline, with a minimum below atmospheric pressure. Alternatively, since this was a buried line, the burning gas generated a pressure outside the pipe greater than that inside. Regardless, it is worth recalling that the split originally formed in the four-inch condensate line on Piper Alpha developed very quickly into a full-bore failure, and it is possible that this too could have been due to an internal explosion.

Underground explosions

Finally, it is only proper to make some reference, albeit briefly, to a problem that has caused many tragic accidents in times past, and has claimed many lives; namely, explosions in underground workings, particularly coal mines.

The explosive in such cases is usually a mixture of hydrocarbon gases with air. The hydrocarbon is primarily methane, and was known to older generations of miners as firedamp. It occurs in pockets whence it may diffuse into mine galleries and passages. At levels close to the earth's surface such a hydrocarbon gas may be produced by the contemporary decomposition of organic matter in ponds or marshy ground, where it is known as marsh gas or carburetted hydrogen. Methane-rich gas is also generated in refuse tips and on sewage farms.

Exceptionally, explosives may be due to a mixture of inflammable dust with air, but these accidents are more commonly associated with aboveground installations such as grain silos.

Methane-air explosions are of the deflagration type, but in the confines of an underground chamber are nevertheless devastating. The ignition source was, in times past, a candle or rushlight. The invention of the safety lamp by Sir Humphrey Davy in 1815 was a major step towards improved safety in coal mines. Electric light is yet safer, but electricity may itself generate sparks, and stringent measures are taken to minimise this hazard. Sparks may also be produced by steel implements for example if they strike rock, but this problem may be avoided by the use of non-sparking tools, usually made from copper-base alloys.

Many of the worst pit accidents (not all of which, of course, were due to explosions) occurred in deep mines. In spite of increased mechanisation, such mines have become less and less able to compete with oil and gas on the one hand and coal exported from more accessible seams in distant countries. So the second half of the twentieth century has seen a sharp decline in the amount of deep-mined coal in industrialised countries. Although not very appealing to those immediately concerned, such developments have had an overall good effect on safety and conditions of work. This is one case where economic pressure combined with changes in technology (the evolution of the bulk carrier for example) have had a beneficial effect. This subject – the way in which technology affects safety – is discussed in Chapter 5.

References

1 Bai Y and Dodd B, 'Tensile and shear instabilities in tensile tests on rods and sheets', *Metals Technol.* 1981 **8** 420–426.
2 Howatson A M, Lund P G and Todd J D, *Engineering Tables and Data* 2nd edition, Chapman & Hall, London, 1991.
3 Skiles J and Campbell H H, 'Why structural steel fractured in the Northridge earthquake', *Weld J* 1994 **73** (11) 66–71.

4 Anon, 'The Alexander L Kielland Accident', Norwegian Public Reports Nov 1981.
5 Pratt W, Lowson M H and Rhodes K T L, *Sea Gem* Enquiry, British Petroleum, London, 1966.
6 Report of the inquiry into the cause of the accident to the drilling rig *Sea Gem.* Cmnd 3409, HM Stationery Office, London, 1967.
7 Lister H C, 'The examination of some suspension links from the 'Sea Gem' drilling platform', SMRE Report Ref A608/543/01, 1966.
8 Report on tiebars T1 to T7, Lloyd's Register of Shipping, R & TA No 7984, 1966.
9 Lonsdale H, *Ammonia Tank Failure – South Africa*, Ammonia Plant Safety, AIChE 1974 pp 126–131.
10 Maddox S J, *Fatigue Strength of Welded Structures*, 2nd edition, Abington Publishing, Cambridge, 1991.
11 Maddox S J (Ed.) *Fatigue of Welded Construction*, The Welding Institute, Cambridge, 1988.
12 Taylor G, 'The formation of a blast wave by a very intense explosion', *Proc Roy Soc London* 1950 **201A** 159–174.
13 Baker W E and Tang M J, *Gas, Dust and Hybrid Explosions*, Elsevier, Amsterdam, 1991.
14 Gugan K, *Vapour Cloud Explosions*, Institute of Chemical Engineers, Riley Park, UK, 1978.
15 Mahoney D G, *Large Property Damage Losses*, 14 ed., Marsh & McLennon, Chicago, 1992.

CHAPTER 4

Natural disasters and earthquake-resistant buildings

This book is concerned with the accidents that befall manmade objects: aeroplanes that fall out of the sky, ships that are wrecked and process plant that explodes or catches fire, for example. It is not intended to encompass that which used, somewhat uncharitably, to be called acts of God, such as tempest, flood and earthquake. These events are due to natural forces over which mankind has no control whatever; nor are they predictable, except to a very limited degree. Furthermore, their power is such as to overwhelm man's feeble defences; or at least this was the case until relatively recent times. However, some technology does now exist for damage limitation in natural disasters. Mostly this is passive, where for example a combination of defences and timely evacuation can reduce the loss of life from flooding to very low levels. In one area, however, great strides have been made. In recent years engineers have developed the skill of designing earthquake-resistant buildings to a high degree. Such buildings are indeed human artefacts, intended to resist the forces of nature, just as a ship is intended to resist storms at sea. Therefore earthquakes and earthquake-resistant structures form part of this chapter. This being so, it is necessary, if only to put matters in perspective, to look very briefly at the incidence of natural disasters, and at how their effects have changed over a period of time.

Natural catastrophes

In May 1994 the United Nations organised a conference on natural disaster reduction in Yokohama, Japan. The incentive for this activity was made clear in the proceedings of the conference,[1] namely, that the cost of such disasters is increasing and this is having a damaging effect on the economies of some member states. For example, in the 1976 earthquake in Guatemala 24 000 people were killed and damage of US$1.1 billion was done. This cost amounted to 18% of the country's annual national income.

Such international relief that may be available is never sufficient to counteract the effect of such a catastrophe.

One positive result of this conference was a study that gave figures for the effects of disasters over a 30-year period (1963–92). These disasters were recorded where they produced one of the following effects:

1 Damage costing more than 1% of the annual gross national product.
2 Significantly affected more than 1% of the total population.
3 Caused 100 or more deaths.

Figure 4.1 shows the incidence of disasters falling into one of these three categories over the 30 year period, while Fig. 4.2–4.4 classify the type of incident falling into the three categories. In all cases there is an increasing trend, which overall amounts to about 6% annually, three times the rate of population growth. It is not suggested that the Earth's climate

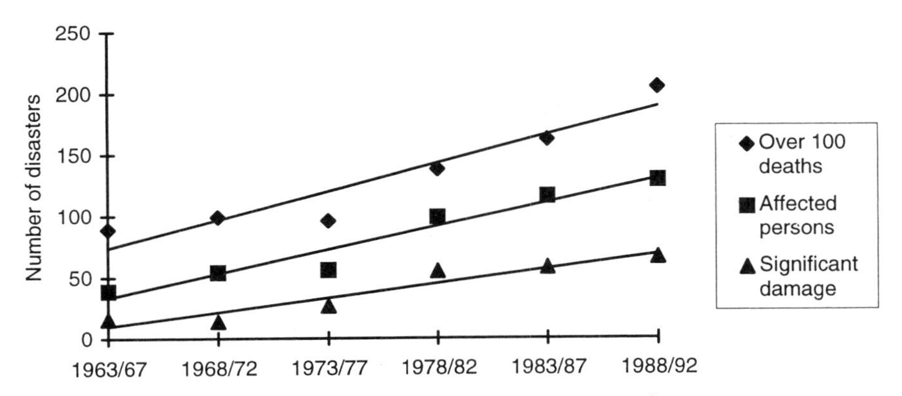

4.1 Incidence of natural disasters, worldwide 1963–92.[1]

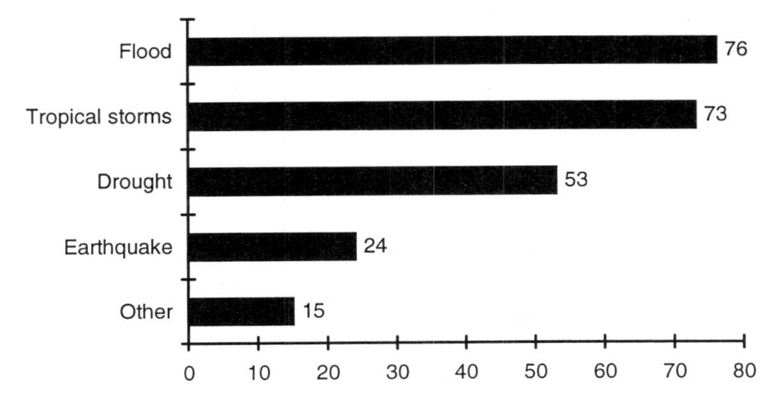

4.2 Numbers and type of natural disaster resulting 'in significant damage'. Other includes storm, fire, volcano, landslides, tidal waves and cold.

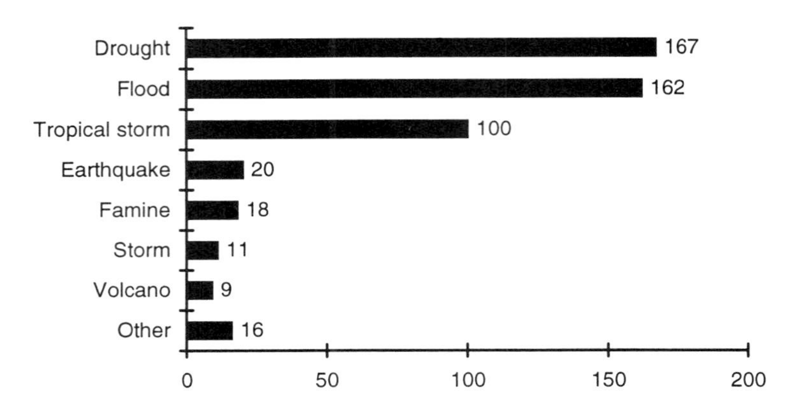

4.3 Number and type of natural disaster affecting more than 1% of population.[1]

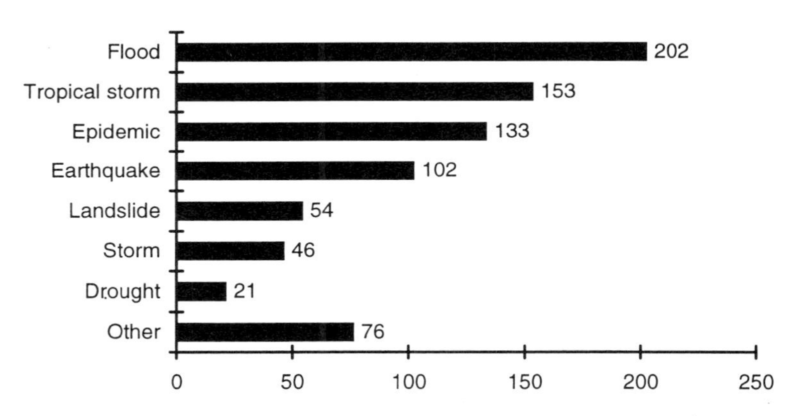

4.4 Number and type of natural disaster resulting in over 100 deaths.[1]

has changed, and it is implied that more people and greater amounts of property are at risk than heretofore.

Overall, the greatest damage is caused by floods, tropical storms, drought and earthquake. During the 30-year period, the number of disasters caused by floods, tropical storms and drought has increased quite sharply, but those caused by earthquakes has remained more or less steady. This conclusion relates to numbers of incidents. A different study, however, indicates that the cost of earthquake damage is increasing (Fig. 4.5). These figures correspond to an annual increase of 20%, which would appear to be excessively high; inflation may possibly account for part of the rise.

One important point to note is that the cost of natural disasters, including earthquakes, amounts to billions of dollars, very much higher

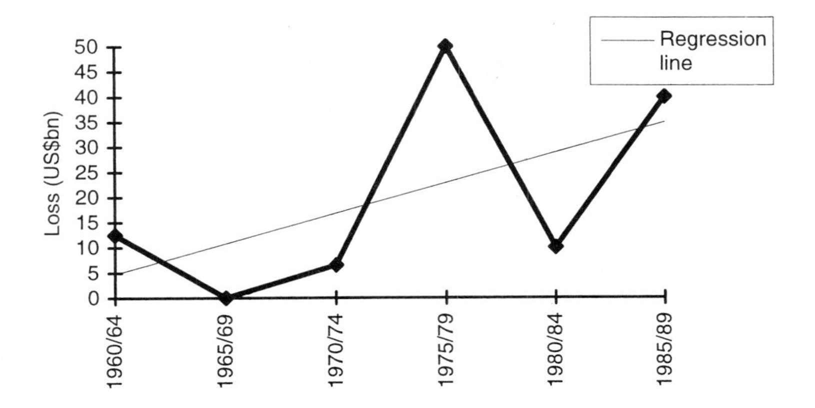

4.5 Economic loss due to earthquakes worldwide 1960–89.[2]

than that due, for example, to vapour cloud explosions in process plant.

The conclusions of the 1994 UN document[1] is, not surprisingly, that more effort needs to be made in investigating the effects of natural disasters. As will be seen in the next section, such efforts are indeed being made, but primarily in developed countries. Less developed countries, which can ill afford the cost of disasters, do not have the ability to put the required measures into place. A very considerable degree of economic development is required to resolve this problem.

Earthquakes

To be involved in an earthquake is a grievous misfortune, not least because they are the most unpredictable of natural phenomena. For many years engineers and seismologists have tried to find a way of predicting the time and place of an earthquake, and occasionally there is an announcement of imminent success. But nothing materialises, and when the next quake occurs everyone is taken by surprise. The effects of an earthquake are equally unpredictable. There is no correlation between the energy of the event, as measured on the Richter scale, and the amount of destruction or loss of life. In part, of course, this is because many earthquakes occur in rural areas. But even when they take place in or near towns, their effects are variable.

The mechanics of the process

According to the theory of plate tectonics the Earth's crust is divided into a number of rigid plates which move relative to each other at rates of 10–

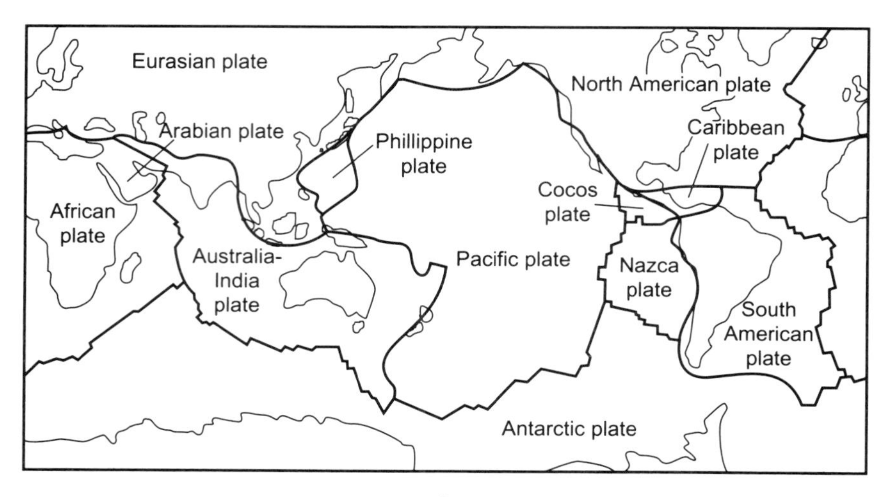

4.6 Boundaries of tectonic plates.[3]

100 mm per annum. Figure 4.6 shows where these plates are thought to be located.[3] At their margins the plates interact in various ways. They may slide laterally against each other, or they may collide, such that one plate rides over the other. Such movement occurs along fault lines or cracks; however, it is not a steady motion and often the crust on either side of a fault locks together until stresses build up and there is a sudden fracture. The result is a large release of energy and elastic waves spread outwards from the fractured region, which is known as the focus. The most damaging earthquakes are caused by those where the focus is between 3 and 9 miles below the surface; these are known as shallow-focus earthquakes. Most seismic events in California are of this type, as was the very destructive 1995 earthquake in Kobe, Japan.

The theory of plate tectonics provides a good explanation of the earth movements around the Pacific ocean, notably in California and Japan. Others, however, have occurred remote from known faults or plate margins. There was a severe earthquake in 1811 which was located in the Mississippi valley. And there was once an earthquake in Birmingham, England, which caused little damage but much alarm among the local population.

Seismic waves

When a sudden fracture occurs, three types of elastic wave are generated: pressure waves, shear waves and surface waves. Pressure waves result from horizontal movement and, at any one point, cause a fluctuating pressure;

they cause little surface movement. Shear waves, on the other hand, result in violent up-and-down and side-to-side motion; these are the waves that cause most of the damage. Surface waves mainly affect the ground surface only; they may also cause both lateral and vertical movement. Pressure waves travel faster than shear waves. Thus, a seismograph will initially show a vibration of relatively small amplitude, but when the shear wave arrives the amplitude increases suddenly. The time interval between these two events provides a measure of the distance between the observation point and the epicentre of the disturbance. The epicentre is the point on the Earth's surface directly above the focus, and if there are observations from two or more stations its location can be obtained.

The magnitude of an earthquake is measured using a scale proposed by the American seismologist Charles Richter in 1935. This number is the logarithm to base 10 of the maximum wave amplitude measured on a standard seismograph situated 100 km from the epicentre. The amplitude is expressed in thousandths of a millimetre. Thus, with an amplitude of one millimetre, the magnitude on the Richter scale would be 3. This number gives an estimate of the total energy of the event.

In the United States and in some other countries earthquake effects are also rated on the modified Mercalli intensity scale. This rating is subjective, and is based on personal reactions to the event, amount of damage to structures and on observations of other physical effects. The intensity is indicated by a Roman numeral, with a maximum of XII.

A more important measurement, so far as destructive effect is concerned, is the acceleration of the ground surface. In principle, this figure could be obtained by analysing the seismographic record to obtain the maximum combination of amplitude and frequency. Suppose the vibration took the form of a sine wave. The maximum acceleration then occurs at the point of maximum amplitude, and is $\delta\omega^2$, where δ is amplitude and ω is frequency. In practice accelerations are best measured on the ground. The acceleration in an earthquake is usually expressed as a multiple of gravitational acceleration, g, which is 32.2 ft/s^2 or 9.81 m/s^2. Thus an acceleration of 16.1 ft/s^2 is expressed as 0.5 g. The highest accelerations so far recorded in California were during the 1994 Northridge earthquake, these being, as noted earlier, 1.8 g horizontally and 1.12 g vertically. A previous maximum had been 1.25 g in San Fernando about 2 miles from the epicentre. Building regulations in Los Angeles require the installation of at least three accelerometers in all new multistorey buildings, so that the urban areas are well covered by these devices.

The other important factor affecting earthquake damage is the duration of the disturbance. The longer the duration, the greater the amount of damage.

The force F on the foundations of a rigid building of mass M during an earthquake causing a ground acceleration a is given by Newton's second law

$$F = Ma = MCg \quad [4.1]$$

where C is the multiplication factor mentioned above. This quantity is known as the seismic coefficient, and is much used in design for earthquake resistance.

Tidal waves

A sub-sea earthquake in which there is a large vertical displacement can produce very large waves capable of travelling considerable distances. Of course they have nothing to do with tides and in more informed circles they are known as tsunani. One such occurred in Hawaii in 1946 as a result of an earthquake in the Aleutian Islands. Waves 50 ft high struck the north-east coast of Hawaii and caused widespread damage and the loss of 173 lives. The effects were felt in Los Angeles, about 1875 miles away.

Liquefaction

Where the ground consists of fine grains, sand or silt saturated with water, earthquake vibration may cause it to liquefy. The foundations of buildings then lose some or all of their support. As a result they may tilt or capsize. In 1964 there was an earthquake of magnitude 7.5 with an epicentre about 35 miles from the city of Niigata, Japan. The ground acceleration in the city was about 0.1 g, not enough to cause serious structural damage, but part of the residential area was built on the sandy soil of a river plain, and there was a high water table. This soil liquefied. Figure 4.8 shows how structures collapsed or tilted during the Luzon earthquake.

In 1976 there was a catastrophic earthquake measuring 7.8 on the Richter scale in north-east China. About a quarter of a million people died in this disaster. The epicentre was near the city of Tangshan, and damage extended as far as Tianjin, 125 miles away. This part of China is low-lying and the water table is high. It was estimated that the soil liquefied over an area of several thousand square miles. There was devastating loss of life in Tangshan and the surrounding area. This was, of course, due to structural damage, not liquefaction. In 1980 a large number of people in Tianjin were still living in brick shacks which had been put up along the roadsides as temporary housing. In rural areas the earthquake formed spouts of mud which spread over the countryside, ruining the harvest.

Earthquake-resistant buildings

There is no such thing as an earthquake-proof building, but a great deal

4.7 A reinforced concrete bridge span that collapsed due to the 1994 earthquake at Northridge, California.

4.8 Titled building in Dagupan, The Philippines. This was the result of liquefaction during the Luzon earthquake of 1990.

can be done to minimise the risk of damage and loss of life by designing buildings that have sufficient strength to resist the forces generated by earthquakes, and sufficient ductility to absorb the energy of the oscillation without cracking or collapse.

The design techniques used for earthquake-resistant buildings have developed from relatively simple formulae used in the early part of the century to the complex and sophisticated codes that are in use today. In Japan earthquake-resistant buildings had been designed and constructed by the time of the 1923 Kwanto earthquake in the Tokyo region, and were reported to have performed quite well. Similar developments came rather later in California and Europe. Japan and the state of California are world leaders in the field, but their approaches to the problems, although similar to start with, have diverged in recent years.

Green[4] gives an excellent account of the way in which the Californian seismic codes have developed. The initial impetus was provided by the Long Beach earthquake in 1933. This caused extensive damage not only at Long Beach but also in the Los Angeles area. The most shocking aspect of this incident, however, was that it caused severe damage to school buildings. The earthquake occurred at 6 am, and it was evident that had it taken place during school hours many children would have been killed. In consequence the state legislature passed acts that, amongst other things, set up requirements for a building code to cover earthquake-resistant structures. These requirements were then incorporated in the building code for the city of Los Angeles.

This code, and its subsequent developments, was based on the notion that as a result of the accelerations associated with an earthquake, the building would be subject to shear forces, and that such forces could be treated as static loads. The shear force is given by

$$F = CW \qquad [4.2]$$

where F is the force, C is the seismic coefficient and W the weight of the structure. No attempt was made to obtain the seismic coefficient from measured accelerations; rather it was (and still is) regarded as an empirical figure derived from experience with earthquakes and their effect on buildings. Initially it was assigned the value of 0.08, except for schools where it was increased to 0.10.

In the case of taller buildings the use of a single coefficient at all levels is not appropriate. The deflection of a multistorey structure is greatest at the top and the acceleration increases correspondingly. Therefore the seismic coefficient must be greater for the upper storeys. In 1943 the code was modified to recognise this fact, and in 1957 it was further modified because in that year Los Angeles city allowed, for the first time, the construction of buildings having more than 13 storeys. The seismic coefficient became

$$C = \frac{4.6\ S}{N + 0.9\ (S - 8)100} \qquad [4.3]$$

where N is the number of storeys above the one under consideration and S is the total number of storeys, except that $S = 13$ for buildings of less than 13 storeys.

At about the time that this formula was incorporated in the city building code, work started on a more comprehensive design technique.

The deflection of tall buildings is affected by their natural frequency of vibration. Generally speaking, the taller the building, the lower the natural frequency, and the deflections and shear forces are correspondingly lower. There are approximate formulae for calculating the vibration period of a building based on its dimensions. So it is possible to make a preliminary assessment using such approximations, and then to carry out a full dynamic analysis to refine the design. Current practice is to carry out such dynamic analyses for all major buildings.

An important innovation at this time was the requirement that the framework of buildings should be ductile. It was recognised that in severe earthquakes the displacements could be large, so the intention was that under such conditions members should bend and not break. At the time it was considered that this condition would be met without reservation by steel-framed construction, but questionably so for reinforced concrete. Nevertheless it was argued that by providing the right sort of reinforcement, concrete could be rendered ductile. 'Ductile' concrete frames were therefore included in the code.

The brittle failure of steel space frame members in the Northridge earthquake of 1994 has called some of these assumptions about steel-framed buildings into question. A good deal of work will be required to resolve this problem.

Actual shear forces

Green[4] has considered the anomalous behaviour of the Holiday Inn building during the San Fernando earthquake. This was a reinforced concrete-framed structure, and it suffered the highest measured acceleration, equal to 0.27 *g* at ground level and 0.4 *g* at roof level. Nevertheless the structural damage was slight, and easily repaired. Calculations were made using the measured accelerations of the shear force at the base of the building. These turned out to be 3½ times the value used in the design of the structure, sufficient, it might be thought, to produce a total collapse. One suggestion (which is by no means universally accepted) is that materials behave differently in high-speed dynamic loading than in a low-strain rate laboratory test. Such differences have already been noted for several cases of dynamic loading earlier in the book. It was also noted that the high acceleration was of a very brief duration so possibly the cracks did not have time to propagate across the complete section of the members.

Japanese developments[5]

Japan is located in one of the most seismically active parts of the world. It lies in a region where the Indonesian and Pacific plates move north along the Eurasian plate. The whole region is under pressure, and a complex system of faults results in frequent tremors and earthquakes. There were no less than four substantial earthquakes between 28 December 1994 and 17 January 1995, the last being a severe shallow-focus quake that caused extensive damage in Kobe and killed over 4000 people.

Japanese engineers have for the most part used the same methods as those described above for designing earthquake-resistant structures. In the late 1960s, however, new ideas were put forward, and some of these, notably base isolation, have been put into practice.

The general approach has been to find means of damping the oscillations of buildings exposed to earthquake shock, and to eliminate the possibility of resonance. The simplest device is base isolation, a system first developed and put into use in New Zealand. This consists of a pad of laminated rubber, on which the foundations of the building rest. Lateral movement of the pad reduces the amount of shake to which the building is exposed, and tends to damp any oscillations. Californian engineers tend to regard this device as too costly to be used other than as a last resort, but of course such costs must be set against the repair or replacement costs following an earthquake.

Other passive devices are sketched in Fig. 4.9. The first of these incorporates a flexible frame at ground level which would reduce shear stresses at the upper floors. Figure 4.9(c) has a damping device consisting of a weight attached to a spring at roof level, and (d) has a brace containing an oil cylinder running across the ground floor frame. A number of other methods of damping using hydraulic systems have been proposed.

The most ambitious project, however, is the intelligent building. A large movable weight is located on the roof, and the motions of this weight are controlled by a computer in such a way as to counter the vibrations due to the earthquake (Fig. 4.10). There are two versions. One employs a remote sensor, which transmits a record of the seismic motions to the computer before it has reached the building. The computer recognises the form of the vibrations and is prepared to initiate the appropriate movement of the weight as soon as the earthquake arrives. The other type has a feedback loop, such that there is a response to counter any motion of the building itself. A number of intelligent earthquake-resistant buildings have been constructed in Japan.

A recent tabulation lists 33 buildings constructed with one or other of the devices illustrated in Fig. 4.9. Of these, 18 were located in Japan. Most were relatively low buildings, with a maximum of eight storeys, of the base isolation type. The remainder employed various types of damping, including the World Trade Centre in New York, which is a steel-frame

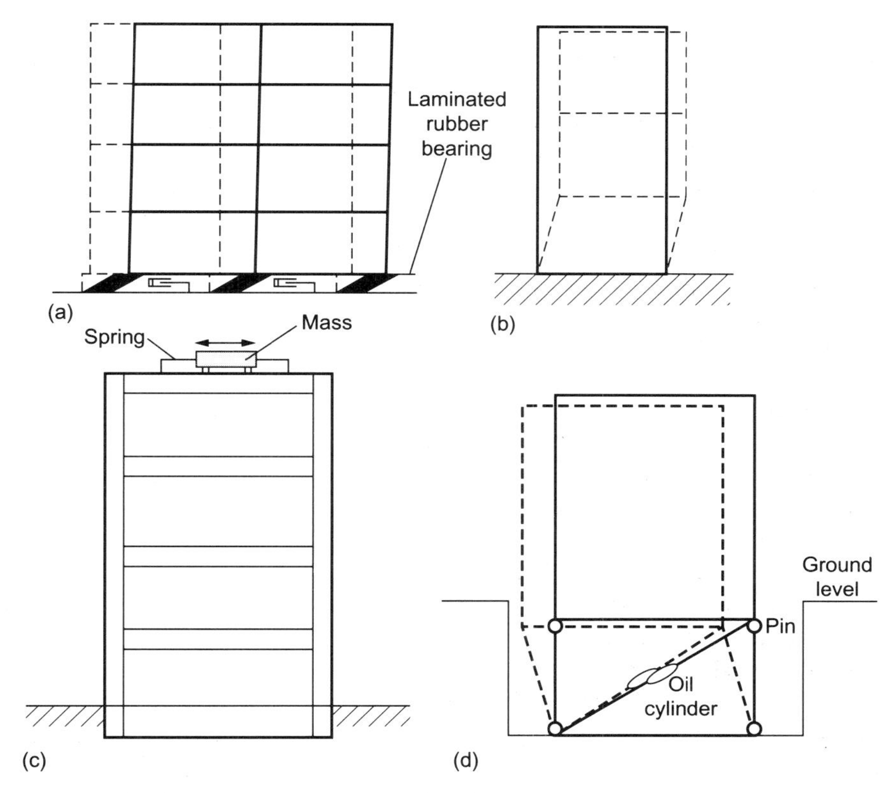

4.9 Japanese proposals for earthquake-resistant buildings: (a) base isolation; (b) flexible first storey; (c) dynamic damper; (d) viscous damper.

building of 110 floors, and which employs visco-elastic dampers; although in this case it is designed to counter wind, not earthquakes.

Do the protective measures work?

In general terms the protective measures do work. Detailed comparisons are lacking. However, during the first half of this century, the average number of fatalities per earthquake in the developed and undeveloped countries of the world were about equal, the numbers in both cases being about 12 000. In the period 1950–92, however, the death rate in the developed countries fell to 1200 per earthquake (equivalent to an annual decrement of about 7%), whilst in the undeveloped world the figure remained at about 12 000.[2] There is little doubt that the improvement in developed countries was due to improved buildings.

The relatively high death toll (for Japan) in the 1995 Kobe earthquake

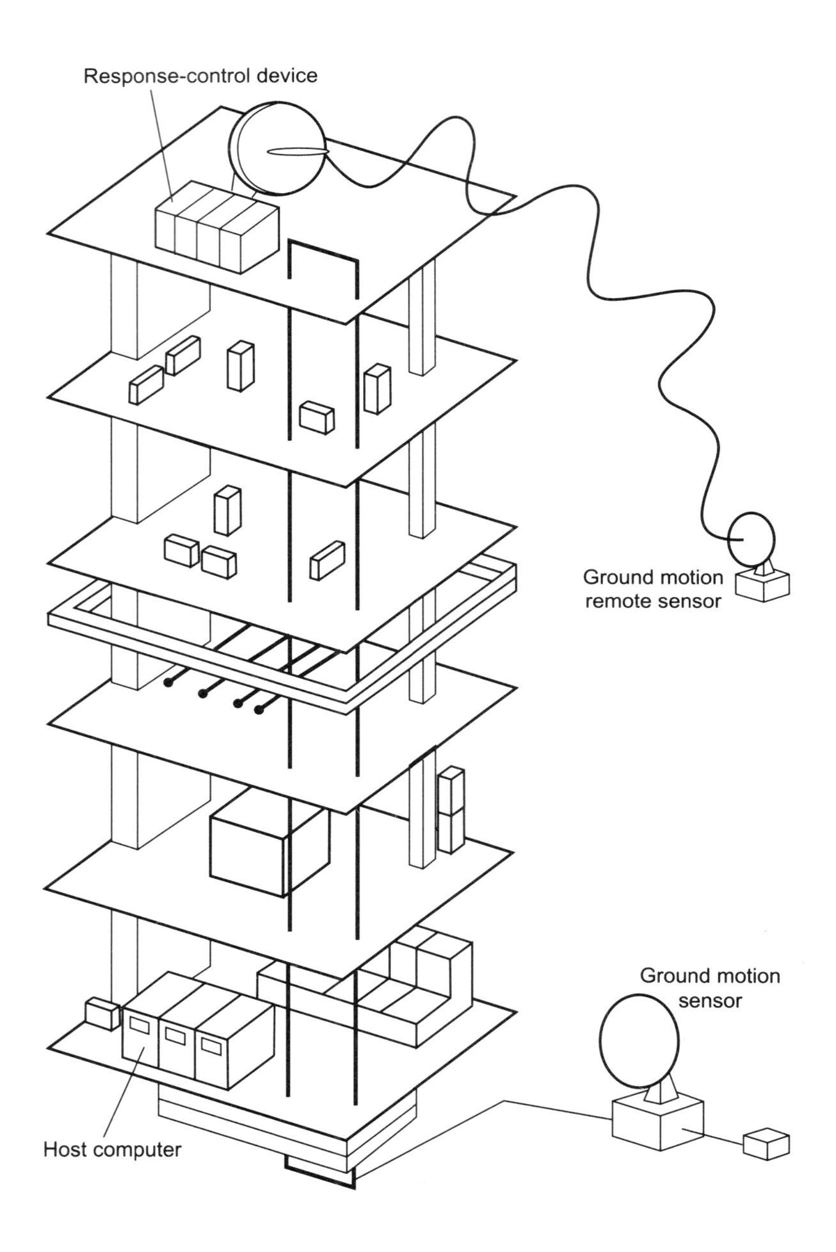

4.10 Earthquake-resistant building with computer-controlled dynamic response; the intelligent building.

(4000+) probably reflects two factors; firstly that it was a shallow-focus earthquake, and secondly that there are, even in the best-regulated city, old buildings that were not designed to the latest codes. However that may be, the overall improvement, by a factor of ten, demonstrates that it is possible to defeat the most fearsome of natural forces.

References

1 Anon, 'Disasters around the world' United Nations conference on natural disaster reduction, Yokohama, 1994.

2 Tucker B E, Erdik M and Hwang C N, *Issues in Urban Earthquake Risk*, Kluwer Academic Publications, Dordrecht, 1994.

3 Gubbins D, *Seismology and plate tectonics*, Cambridge University Press, Cambridge, 1990.

4 Green N B, *Earthquake Resistant Building Design and Construction*, Elsevier, New York, 1987.

5 Anon, *Technological Development of Earthquake Resistant Structures*, Japan Building Centre (English trans.), Balkema, A A, Rotterdam, 1993.

CHAPTER 5

How technological change affects safety

A good deal of evidence has been presented earlier in this book to show that for most branches of industry and in transport, the fatality rate and the rate of loss of equipment is falling with time. Since the manufacturing techniques generally improve with time, it might be deduced that an improvement in technology will result in increased safety.

Unfortunately this is not always the case. For example, the increased use of steam power in industry during the late nineteenth century was accompanied by a large increase in the number of boiler explosions, with a corresponding rise in the number of deaths. Further improvements in boiler technology improved the situation, but the problem was really solved by the introduction of electricity as the base power source.

Likewise, the large increase in the number of motor vehicles that occurred during the first half of the twentieth century resulted in a rapid increase of deaths and injuries due to road accidents. This became a matter of great public concern. The Monday morning newspapers dismally recited the figures for the previous weekend's death roll, and the government of the day promoted road safety campaigns. In recent years, however, the situation has been brought under control. The compulsory use of seat belts and improvement in car design have resulted in a marked reduction in the annual number of road accident fatalities, and this trend is expected to continue. A similar pattern of rise and fall was noted for aircraft losses after the introduction of new types, although this occurred over a much shorter period of time and had little impact on the overall statistics, which show a continued reduction in the annual percentage loss of aircraft. Thus, the introduction and widespread use of certain types of equipment has in the past led to a temporary reduction in safety, and the same problem could occur in the future.

The case of hydrocarbon processing plant is quite different. Here there has been an increase in the rate of loss of plant in recent years, largely due to an increase in the number and severity of vapour cloud explosions. In a later section the change in the process fluids is discussed; from the relatively non-hazardous kerosene and gas oil produced at the end of the nineteenth century to the potentially explosive substances such as ethylene and high-pressure hydrogen that are handled a century later. This change has been dictated by market demands. Changes in industrial requirements have made hydrocarbon processing inherently more risky, and the means of counteracting this increased risk have not yet been found.

In almost all other engineering activities technological change has been beneficial, and the details are set out below. But first it is necessary to consider one apparently intractable problem; the relatively high and apparently unchangeable fatality rate in the construction industry.

Safety on construction sites

In Chapter 1 it was shown that in some areas of human activity, the proportional loss of life and equipment diminished from year to year whereas in others, notably offshore mobile operations and construction, there has been no such improvement or at best it has been slow. In general, those industries where technological change had a predominant effect on safety improved the most. Air transport, for example, is a field in which technical improvement has proceeded at a great pace throughout the twentieth century, and the rates of loss of life and aircraft have diminished accordingly. In the construction industry, on the other hand, fatalities have fallen at such a low rate that a single large accident could reverse the trend. It is self-evident that construction work will be less affected by technological change because many of the accidents are those that occur in everyday life: being struck by a vehicle, falling from a height and so forth. And it is just as easy to fall off a ladder today as it was yesterday. There have been improvements of course. Vehicles have become safer and more reliable. To some extent, however, this improvement has been offset by increased speed; some earthmoving equipment now runs at a quite alarming pace. So any improvements must rely heavily on positive efforts by construction organisations to improve safety standards on job sites. This book is not concerned with such matters, important though they may be; we are dealing here primarily with the technical background to catastrophes. Nevertheless it is worth looking at the record of one particular company, if only to demonstrate that persistently high casualty rates are not inevitable, and that they can be brought under control.

The company in question is M W Kellogg Company, a US-based contractor that operates worldwide in the design and construction of oil and petrochemical plant. In recent years this company has set up a safety programme for its construction work. All aspects of the problem are covered, including government requirements, which are increasingly stringent in all developed countries, co-operation with labour unions, design to improve constructability, making all managers responsible for safety, scrutinising records, training and communication. The results are to be seen in Fig. 5.1. In the space of a few years, the recordable incident rate has been reduced from a level that was above the national average to one that is substantially lower, and this lower rate has been yet further reduced. Over the same period the US average accident rate for the construction industry showed only a modest fall, very like that recorded for fatal accidents in Britain (Fig. 1.8). So given the will, and a sound approach to the problem, it is possible to reduce accident rates without any radical change in technology.

Developments in steelmaking

A number of the major catastrophes described in previous chapters were due almost entirely to the use of steel whose quality was inadequate for the job. Examples are the *Sea Gem*, the Liberty ships and the *Alexander L Keilland*. This is not to say that any of the designers concerned specified or used defective material knowingly; they were simply unaware of the potential problem at the time of the original construction. So it is of interest to see how far the intrinsic quality of metals, particularly steel, has improved and to consider how far this may affect safety.

During the twentieth century virtually all steel has been produced by the indirect process; that is to say, iron ore is first reduced in the blast furnace to produce pig iron containing about 4% carbon, then the carbon content is reduced to less than 1% in a steel-making furnace or converter. The direct reduction of iron ore using hydrogen, for example, is technically possible and has been practised on a limited scale, but it requires a high purity ore, and is not generally feasible.

Before World War II a substantial proportion of the steel made in continental Europe was produced in basic Bessemer, or Thomas, converters (Fig. 5.2). This process was used primarily for iron that had been made from basic ores rich in phosphorus. The basic lining reacts with the phosphorus to produce a slag which can be used as an agricultural fertiliser. The converter was rotated so that its long axis was horizontal; liquid iron was poured in, it was then placed in a vertical position and at the same time air was blown through the nozzles or tuyeres in the base. In this way the carbon could be reduced to the required level in about 15 minutes. The productivity of the Thomas process was high, but the high nitrogen content of the steel, picked up from the air injection, was undesirable and a number of modifications were introduced, such as blowing with a steam–oxygen mixture.

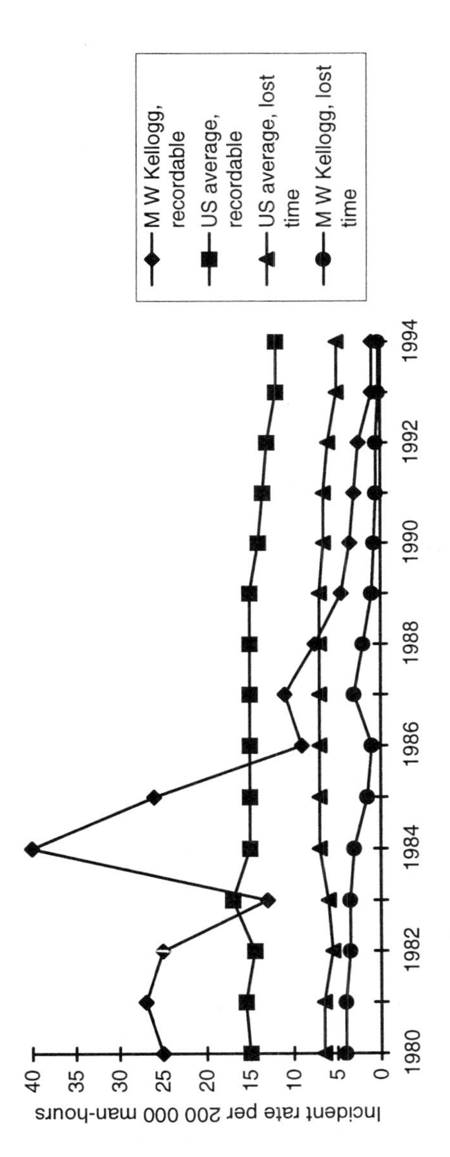

5.1 The effect of a safety management programme on lost time, injury and death in the construction of process plant.

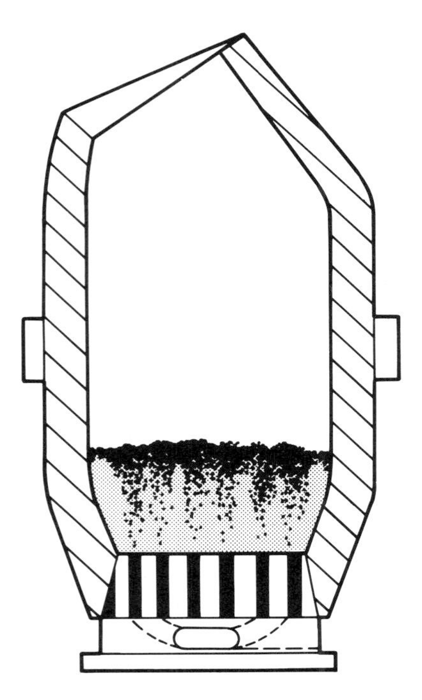

5.2 Bessemer converter. After the blow, the convector body is rotated through a little more than 90° to discharge the refined metal.

During the same period most British and US steel was made in open-hearth furnaces. These were large horizontal furnaces capable of handling 200–400 tons of iron. The bath was relatively shallow and heated by gas that flowed across the surface of the melt and then down into recuperators below. The carbon was oxidised by additions of iron ore, and in later years also by injecting oxygen through a long tubular lance. Productivity was lower than with the converter but it was possible to include scrap steel as part of the charge.

The third important steelmaking furnace at that time was the electric arc furnace. This was used primarily for remelting scrap.

Liquid metal from these furnaces was poured into ladles and thence into ingot moulds. Before pouring, a deoxidant could be added to the ladle. In continental Europe the steel was either fully killed or rimmed. In rimming steel, there was no added deoxidant and during cooling prior to solidification, carbon in the steel reacted with oxygen to produce an effervescence of carbon monoxide in the cooler parts near the mould surface. As a result the outer surface of the ingot was almost pure iron, but the interior was porous with impurities concentrated along the axis.

Because of the porosity there was no shrinkage cavity at the top of the ingot.

In fully killed steel practice, aluminium was added in the ladle; this combined preferentially with the oxygen so there was no porosity and because of the contraction during solidification, a relatively large shrinkage cavity formed at the top. UK and US steelworks favoured a semi-killed practice, where enough deoxidant (silicon and manganese) was added to prevent the rimming action, but allowing some porosity to form so as to minimise the amount of shrinkage. The economics of steelmaking was much affected by the amount of ingot that needed to be cut off to avoid defects in the rolled plate, and various other devices, such as heating or insulating the top of the mould, or pouring through a sprue into the bottom of the mould, were used to reduce waste.

After cropping the defective portion of the ingot, if any, it was reheated and rolled in a reversing mill. This consisted of a set of rolls with a roller-bed on either side. The slab was manipulated by pushers to and fro through the rolls until it had reached the required thickness. This operation was usually conducted in two stages. The ingot was reduced to a thick slab in the slabbing mill, and then this slab was rolled down to plate thickness in a second mill. Alternatively, in the second stage the metal was processed continuously through a series of rollers (the continuous strip mill) and finally wound into a coil. It was expected that any pores or other discontinuities in the ingot would weld together during the rolling process but sometimes this did not happen, and laminations in the finished plate were not uncommon. Typically, the steel so produced contained about 0.25% carbon, 0.04% sulphur and 0.04 phosphorus. It was tested for yield and ultimate tensile strength but only exceptionally for impact strength.

In 1948 the Austrian steel industry faced some special problems.[1] Iron made from locally available ore was not suitable for processing in basic Bessemer converters, so steel was made in open-hearth and electric furnaces. A substantial proportion of the output consisted of special high quality steel that was exported, and the supply of local scrap was small. There was a need to increase production, and it was decided to develop a converter process using oxygen as the agent for removing carbon and other impurities.

Bessemer had patented the use of oxygen for bottom-blown, side-blown and top-blown converters in 1856, so the idea was not new. However, in the nineteenth century adequate supplies of oxygen were not available. Also, as it proved later, oxygen blowing results in high local temperatures, which destroyed the nozzles or tuyeres. It was thought necessary to blow the gas through the liquid iron in order to get sufficient circulation. So it was decided to try top-blowing, bringing the tip of the nozzle either close to or just below the metal surface. Legend has it that during one of the test runs the lance broke well above the metal surface but the final result was not affected. So this arrangement became the norm.

The oxygen blow results in a violent bubbling action, which continues until the carbon content falls to about 0.05%. The liquid metal and slag undoubtedly circulate, but the cause of this circulation is unknown. Regardless, a new steelmaking process had been developed, and by November 1952 the first unit was in operation. In Austria the process was called 'L-D', from the names of the towns, Linz and Donawitz, where the first tests were made. Elsewhere it is usually called the 'basic oxygen process'.

The metallurgical advantages were clear from the start. Impurities, particularly sulphur and phosphorus, were reduced to a low level without introducing nitrogen; indeed the nitrogen content of the original iron fell. Secondly, the operating time was short. Figure 5.3 shows refining curves for basic oxygen, basic Bessemer, acid Bessemer and open-hearth furnaces which illustrate this point.

The improvement in productivity was even more astonishing. By the early 1980s the vessel capacity had been increased to 400 tons, giving an output of 600 tons/hour, about 15 times that of a 400-ton open-hearth furnace. Naturally, the process was adopted worldwide, and now most of the world's steel is made in basic oxygen converters.

One effect of the large potential output of the oxygen converter has been to make continuous casting a commercial possibility. Figure 5.4 shows the layout of a typical continuous casting machine. Liquid metal pours from a tundish into a water-cooled collar where a solidified skin is formed. The partially solidified strand then moves in an arc from vertical to horizontal, guided by rollers, and is eventually cut into slabs by flying shears. The advantages of such a method will be evident; the waste due to cropping ingots is eliminated, and the slabbing mill is no longer required. There are also technical advantages. The steel is all aluminium-killed, and there is no porosity and no shrinkage cavities. Therefore the risk of lamination is much reduced. Laminations in themselves have rarely caused any failures but, as seen in Chapter 2, the repair of discontinuities close to welds (which is required by some construction codes) can be hazardous.

There have been other improvements affecting steel quality. In an integrated steelworks, iron from the blast furnace is conveyed to the converter in a torpedo car, which is an elongated ladle running on rails. Sulphur may be removed in the torpedo car (or in ladles) by treatment with magnesium, calcium or lime, whilst phosphorus can be removed using a basic oxidising slag. By such means it is possible to produce an iron with 4% carbon, 0.005% sulphur and 0.015% phosphorus. These values

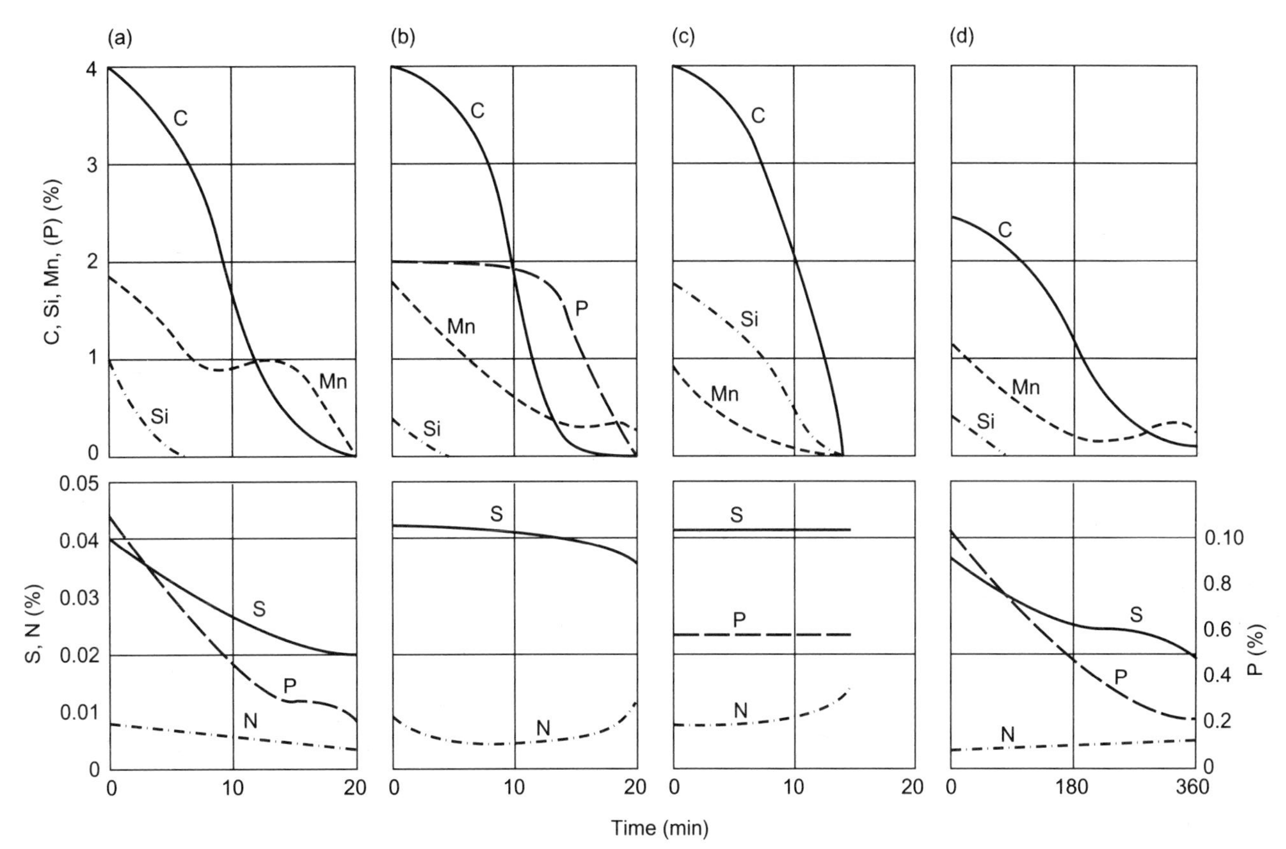

5.3 Refining curves for various steelmaking processes: (a) basic oxygen; (b) basic Bessemer; (c) acid Bessemer; (d) open-hearth.

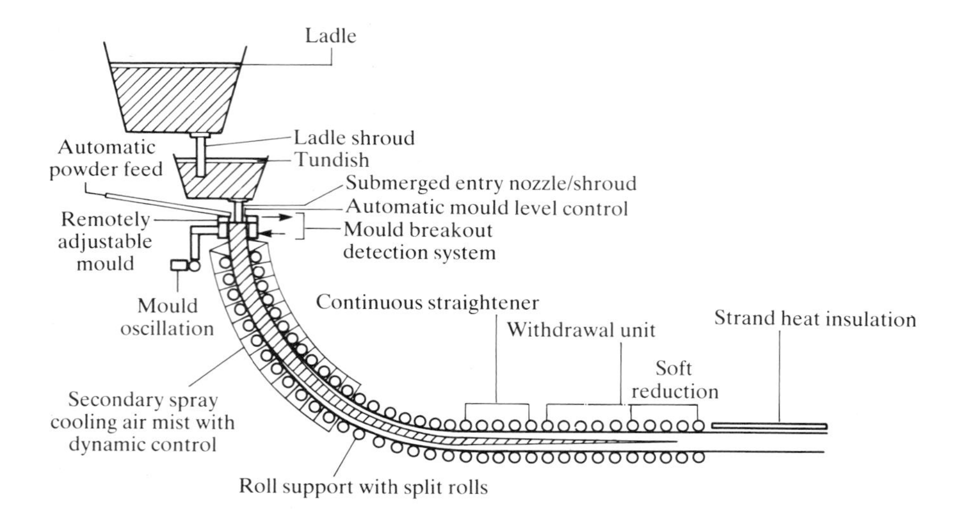

5.4 Continuous slab casting machine.

may be lowered still more by treatment after the oxygen converter. So it has become commercially possible to make a clean steel free from the laminar defects which reduce the through-thickness ductility of plates.

Finally, there have been advances in rolling mill practice. The objective of these developments has been to produce a steel that is fine grained in the as-rolled condition. Fine grain is normally associated with good notch-ductility and a low ductile–brittle transition temperature, combined with increased yield strength. In hot rolling, steel is subject to deformation at high temperature, and by controlling this operation it is possible to arrive at the required mechanical properties.

The first steps in this direction were taken in some European mills in the late 1950s, where rolling was continued below the austenite–ferrite transition temperature. This did indeed have the effect of producing a fine grain, but it also developed a banded structure and elongated the sulphides, such that the material was subject to a laminar weakness. Subsequent developments have been directed towards controlling the operation at temperatures above the transition. The operation is known as 'thermomechanically controlled rolling' and it has many permutations and combinations. To take an example: when rolling is carried out at temperatures above 950 °C, the austenite grains are broken down by the mechanical treatment, and subsequently new grains are formed and start to grow. By allowing the optimum time for recrystallisation after the final rolling pass in this temperature range, a suitably fine austenite grain size is obtained, and eventually this structure transforms to give a fine ferrite grain size. In practice, rolling continues at temperatures below 950 °C and above the transition temperature to obtain further refinement. Figure 5.5 and Table 5.1 shows some of the variations that have been used in the production of the Japanese steel HT50. This is a high tensile steel (ultimate strength about 500 N/mn^2) used for welded constructions such as large storage spheres. The transition temperatures are very low indeed (Table 5.1). HT50 is a special steel, but the technique is applicable also to lower-tensile grades.

It will be seen in Table 5.1 that small amounts of the elements niobium, titanium and vanadium have been added to some of the steels. These are known as microalloying additions, and they act in various ways to refine grain and increase strength. Titanium in combination with nitrogen, restricts the growth of austenite grains in the temperature range 1050–1100 °C. Niobium raises the recrystallisation temperature of austenite. Both these additions have the effect of reducing the grain size of the steel. Vanadium has only a modest grain-refining effect but increases the tensile properties by precipitation-hardening at temperatures below about 700 °C.

The overall result of such developments is that steelmakers can produce material of higher tensile strength and lower ductile–brittle transition temperatures at a relatively modest increase in cost. They have also resulted in an improvement in the quality of ordinary grades of carbon steel. Figure 3.23 shows a comparison between steel made in the 1960s with that made in the late 1970s. There is a reduction in the 27 J Charpy transition temperature of about 50 °C. Since that time quality has improved still further.

Aluminium

The problems that are so dominant in the use of steel, such as the embrittling effects of the common impurities, sulphur and phosphorus, and the transition to a brittle form at low temperature, do not exist in the case of aluminium. In order to extract aluminium, the oxide ore bauxite is dissolved in molten cryolite (sodium aluminium fluoride), and the solution is electrolysed. This process will only operate satisfactorily if the bauxite is of high purity. So the total impurity content does not exceed 0.5%, and the contaminants, mostly iron and silicon, do not have any embrittling effect. Thus there was not much to be done by way of improving the method of extraction. Most of the significant alloy development took place in the early days: the patents for the electrolyte extraction process were granted in 1886, and the age-hardening alloy Duralumin was invented by Alfred Wilm in 1909. Duralumin was a 3½% copper ½% magnesium

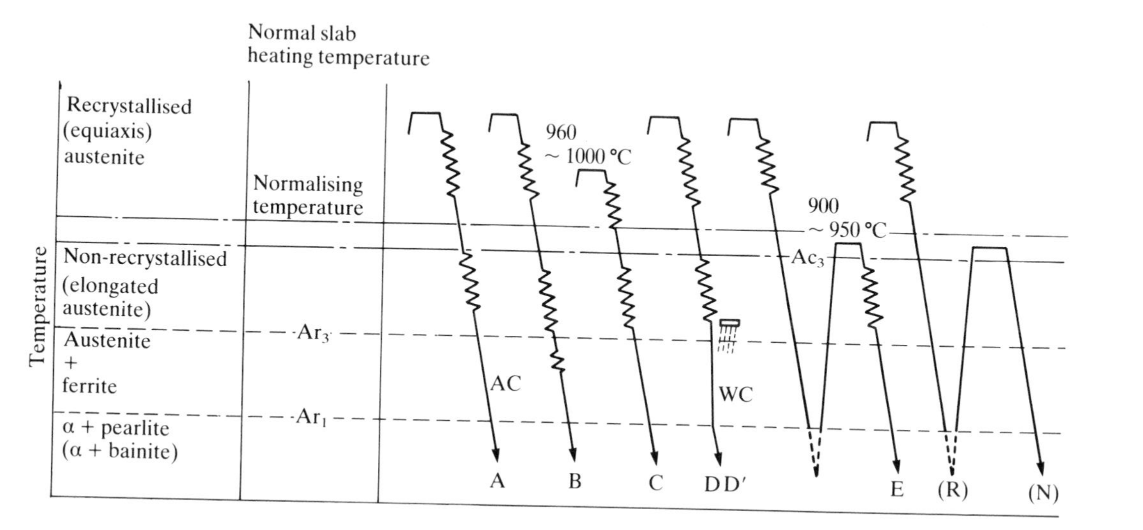

5.5 Controlled rolling of Japanese HT50 steel: thermal cycle. AC, air-cooled; WC, water cooled; A, controlled rolling in γ region; B, controlled rolling also in γ and δ regions (separation); C, controlled rolling, low slab temperature and Ca-Ti; D, controlled rolling and water cooling; E, controlled rolling with reheating just over Ac_3; (R), as-rolled; (N), normalising.

Table 5.1 Controlled rolling of Japanese HT50 steel: mechanical properties

Rolling programme	Chemical composition (%)									CE	YP	UTS	Elongation	VE	50%	Drop-	Grain
	C	Si	Mn	P	S	Nb	V	Ti	Ni	(IIW)	(N/mm^2)	(N/mm^2)	(%)	−40 °C (J)	VTrs (°C)	weight (°C)	size ASTM
A	0.12	0.32	1.36	0.015	0.006	0.019	–	0.013	–	0.36	432	504	29	245	−91	−50	7–9
B	0.13	0.36	1.42	0.02	0.003	–	–	–	–	0.37	368	526	31	139	−92	−95	8–11
C	0.06	0.22	1.32	0.01	0.001	–	0.04	0.01	0.26	0.31	392	471	40	294	−125	−100	11–12
D	0.12	0.27	1.13	0.016	0.004	–	–	–	–	0.31	362	493	30	288	−71	−45	7–9
E	0.05	0.31	1.39	0.019	0.004	0.03	0.05	–	–	0.29	385	469	34	110	−97	−100	11–12
As-rolled	0.13	0.25	1.37	0.013	0.007		REM	0.035		0.36	420	530	22	200	−42	−10	4–6
Normalised											360	500	27	270	−80	−35	6–7

CE = carbon equivalent, YP = yield point, UTS = ultimate tensile strength, VE = impact energy, 50% VTrs = temperature for 50% fibrous fracture, REM = rare earth metals.

alloy which, with minor additions, was used for constructing the Comet aircraft and is still in use today as alloy 2024. The fact that Comets suffered catastrophic brittle failures was not due to any deficiency of the material. So far as is known, there have been no subsequent failures of this type, and the fracture toughness of aluminium alloys is not a matter for concern.

The introduction of the argon-shielded tungsten arc welding of aluminium just after World War II was a major advance. It did not affect the age hardening alloys used in aircraft, because these cannot be joined by fusion welding. It was, however, applicable to the non-age hardening aluminium–magnesium alloys. These alloys are currently used for very large structures, notably offshore accommodation and ship superstructures.

Aluminium–magnesium alloys are benign materials, not subject to embrittlement or weld cracking or corrosion other than in extreme conditions. There has, therefore, been little need for improvement in material quality for either the heat-treatable or the non-heat-treatable aluminium alloys.

The effect of improved steel quality

It is unlikely that any of the catastrophic failure of steel structures described in Chapters 2 and 3 would have occurred had steel of 1990s quality been used; except perhaps for the case of the *Sea Gem*, where the damaging effect of surface weld runs on the tiebars could have been an overriding factor. This does not, of course, mean that the danger of such failures has been eliminated. Brittle fracture of steel having good notch-ductility is still possible under impact loading, and where a fatigue crack has been allowed to grow to an excessive length. Fatigue failure remains no more but no less a threat than in times past. It is not possible to produce a metal that will be immune to fatigue cracking. Avoiding such failures is not a materials problem, but rather a question of careful attention to detail at the design stage, combined with inspection during service. In this connection it is profitable to apply egalitarian principles. Designers are apt to designate particular areas in a structure – the nodes in the case of an offshore tubular construction, for example – as being 'critical'. Under fatigue loading conditions, however, all loaded joints, and particularly welded joints, must be given equal consideration.

It should be noted that good notch-ductility does not make a structure immune to sudden collapse. The *Alexander L Kielland* was made of good quality, notch-ductile steel. However, had this not been so, and had the steel been as brittle as that of the *Sea Gem*, it would have made little difference to the outcome. The important point is to use all means to avoid the type of failure that could give rise to such shock loads.

Aircraft

From the time the first aircraft was built, safety has been a priority for designers; nobody is going to fly in an unsafe machine if a safe one is available. In the early days, however, pilots did not have the same preoccupation. Many were trained by air forces during the two world wars, and a fighter pilot who thought of safety first would not have been much good at his job. So even airline pilots between the wars expected to cut something of a dash. Such youthful exuberance is no longer appropriate in a computer-controlled age. To quote one prophet:

> The cockpit of the future will become in effect the work and command station of a 'flight management' crew who have available the full range of information required for decision-making, the interfaces required with the rest of the world and the controls needed to implement their decisions and fly the aircraft.

A far cry from Flight Officer Biggles!

Before looking in detail at how aircraft have developed, it is worth recalling the history of the airships, and in particular that of the disastrous R 101.

Airships, in hindsight, never had much chance of success. Their basic defect was that even using the lightest gas, hydrogen, the achievable lift was relatively small, and it diminished with height, such that the machines were forced to travel at about 1000–2000 ft, an altitude at which they were highly vulnerable to weather. The optimum cruising speed was low, typically 50–60 mph. Between the two world wars, though, these problems were outweighed by the potential advantage of range, which was measured in thousands of miles, and a passenger capacity that was potentially many times that of contemporary aircraft. Thus, it was possible to think about providing round-the-world passenger services with comfort and leg-room, and with journey times reduced from weeks to days.

Airships were flown in France in the middle of the nineteenth century, but the first practical passenger-carrying craft was produced by Count Ferdinand von Zeppelin, who developed the idea of setting a row of spherical gas bags in a tubular steel framework, the whole being powered by internal combustion engines driving propellers, and suspended below the framework. Prior to and immediately after World War I, a German company ran a large number of pleasure flights in Zeppelins without a single casualty but during the war it was otherwise. Zeppelins were employed to bomb London but the losses were so high that their use was abandoned in 1917 in favour of heavier-than-air craft. In the 1920s and 1930s Zeppelins operated on long-distance flights, including a scheduled service between Europe and South America. In the early years of the twentieth century, therefore, airships seemed to have a bright future.

The British airships

In England airships had a shaky start. The first was designed by Vickers Limited on the lines of a light submarine. Unfortunately it turned out to be over-heavy, and to lighten it, the keel was removed. As a result it broke its back while being towed out into open water. Subsequent developments were all land-based, the design relying heavily on information obtained, in one way or another, about Zeppelins. There was a success, the R 34, which flew to New York and back in 1919, and a catastrophe, the R 38, which broke up in the air during a demonstration flight. Then, in 1924, it was decided to proceed with the construction of two large ships, the R 100 and R 101, that would pioneer regular passenger services to India.

An initial step was the construction of mooring masts. These were towers about 200 ft in height to which the aircraft was attached. Figure 5.6 is a sketch of the layout at the top of the tower. The airship was towed to the mast by a cable powered by a stationary engine, and was steadied by

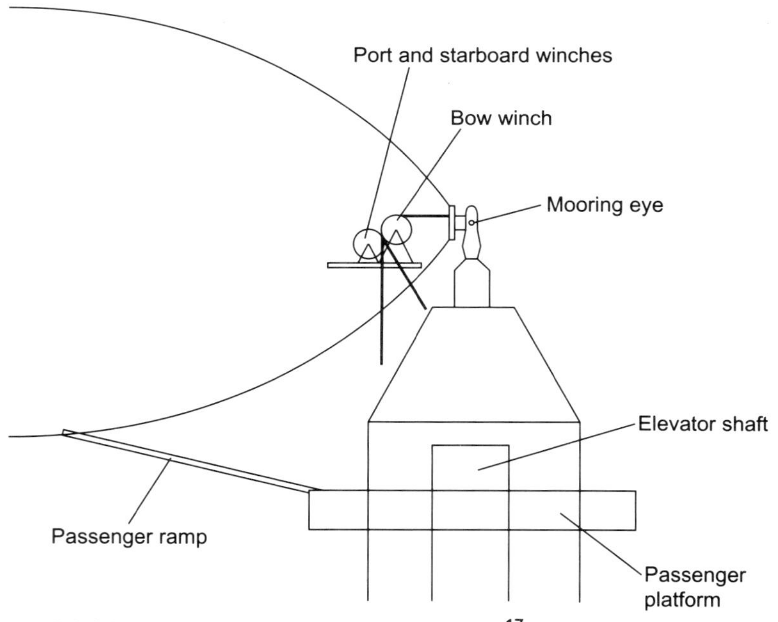

5.6 Schematic diagram of airship mooring mast.[17]

side guys. Passengers came up by lift and boarded by means of the ramp shown.

Mooring masts were built at the Royal Aircraft Works in Cardington, Bedford, England, in Ottawa, Canada, Karachi (then India) and Ismalia, Egypt, which was the intended intermediate stop on the India route. Between flights the airships were housed in sheds, into and out of which they were moved by large gangs of men. If the crew were sufficiently skilled, a mooring mast was not essential. When the Graf Zeppelin visited Cardington it was manœuvred within a few feet of the ground and held by guys whilst passengers disembarked and embarked using a gangway.

The Ministry specifications for the R 100 and R 101 required that the gasbag capacity be five million cubic feet, and that the ships be capable of carrying 100 passengers to India with one refuelling stop in Egypt. They were to be capable of flying at 70 miles per hour and to have a speed of at least 63 miles per hour average over a 48-hour period. One ship was to be built under contract by Vickers Ltd, and the other by the government establishment at Cardington. This inept arrangement established two design and construction teams, one of which (Vickers) was working for and reporting to the other (the government establishment at Cardington); inevitably there was friction. It happened that one of the designers of the Vickers airship, the R 101, was the novelist Nevil Shute (in real life Nevil Shute Norway) and he has left a vivid account of this rivalry.[2] At a later date Peter Masefield, who advised the British Government about the possibility of airship development after World War II wrote a very detailed history of the other craft, the R 101.[3] The subject has therefore been well documented.

A major handicap to the whole project was the belief, firmly held by engineers in the Air Ministry, that petrol engines were unsafe under tropical conditions. It was thought that the low flash point of gasoline meant that they would explode or catch fire in a hot climate. Therefore the Cardington team opted for diesel engines, and as a result suffered a weight penalty of 7 tons. Since Vickers used petrol engines, the R 100 was considered unsuitable for the India route and was scheduled to go to Canada instead.

In other respects the general arrangement of the two ships was rather similar. Both had passenger and crew accommodation inboard, located just above the control car. On R 100 the accommodation was on three decks, on R 101 it was on two. The outline of R 101 is shown in Fig. 5.7, and details of the accommodation are given in Fig. 5.8.

There was however a very important difference. Table 5.2 shows the available lift of the two ships on the date of their first flight (both in 1929) compared with 1924 Ministry requirements. R 100 was 10 tons short, but

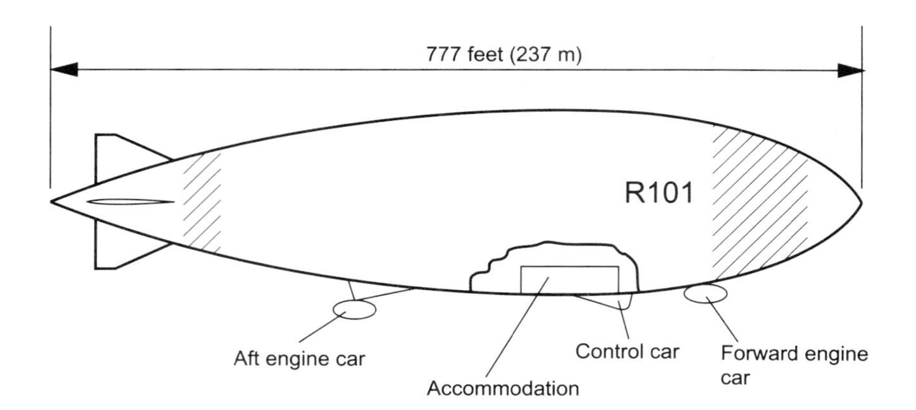

5.7 Outline sketch of airship R 101. The cross-hatching indicates areas of the outer cover that were not replaced during the 1930 refit.

it could still have carried (in theory at least) about 40 passengers to India, whereas R 101 was not capable of making the trip even without passengers. So it was decided that R 101 would be modified by inserting two extra bags, thereby increasing the lift by about 10 tons. R 100 meantime carried out its initial test flights, which were passed satisfactorily. There were, however, problems with the outer cover which required modification.

The outer cover was an unending problem with airships. It consisted of cotton or linen cloth which was stretched over the framework of girders, and then treated with dope. This substance, also used at that time for heavier-than-air machines, consisted of a solution of cellulose derivatives such as cellulose acetate dissolved in a volatile liquid, and it had the effect, it was hoped, of increasing the strength of the cloth. However, tears were quite frequent and repairs were required during and between flights. Covers often leaked in a rainstorm and the gasbags became soaked. Alternative materials were Duralumin or stainless steel, but both were considered to be too heavy. No solution had been found to this problem when airships were finally grounded.

By early 1930 political problems had become acute. The programme was in its sixth year and there were no visible results, except for a round-Britain tour by R 100. An Imperial Conference was due to begin in September 1930 and it was planned that the Secretary of State for air, Lord Thomson, should present an opening paper on the progress of imperial communications. It was intended that in October the initial flight to India would be made, with Lord Thomson on board, and that he would return in triumph to present a follow-up paper to the conference.

Windows
Promenade deck
Cabins
Lounge
Dining room
Promenade deck
Windows

Upper deck

Crew cabins
Cabins
Switch room
Chart room
Crew room
Smoking room
Control car
WT cabin
Galley
Crew cabins
Cabins
Pantry

Lower deck

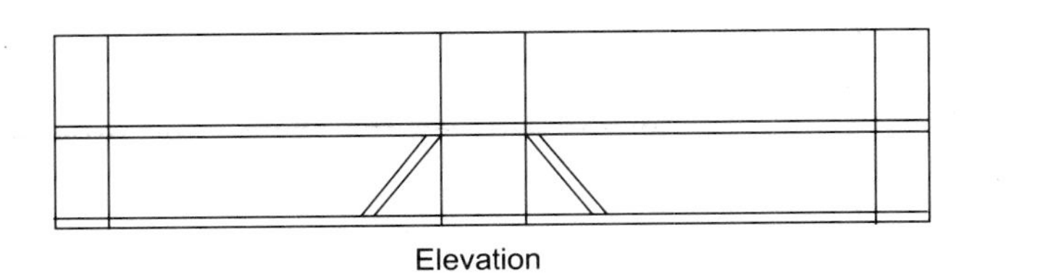

Elevation

5.8 Interior layout of airship R 101.

Table 5.2 Airship weights: UK 1924 programme

Item	Air ministry specification 1924	Airship at the time of first flight	
		R 100	R 101
Volumetric capacity, cubic feet	5 million	5.156 million	4.894 million
Standard gross lift, tons	151.8	156.5	148.6
Fixed weight, tons	90.0	105.5	113.6
Disposable lift, tons	61.8	51.0	35.0
Ship prepared for service, with crew, stores and equipment, tons	110.0	125.5	133.6
Lift available for fuel and payload, tons	41.8	31.0	15.0
Fuel for journey to Egypt, tons	25.0	25.0	25.0
Lift available for payload, tons	16.8	6.0	Nil
Allowance for passengers (350 lb (159 kg) each), tons (number)	15.6 (100)	6.0 (38)	Nil
Allowance for mail, etc, tons	1.2	Nil	Nil

As a start R 100 (Fig. 5.9) was, after many delays, despatched to Canada at the end of July. Major G H Scott, a flyer of the old, press-on-regardless, school was in charge. The trip was a success in that the airship returned to Cardington in one piece. However, in Canada they had encountered a thunderstorm, and instead of avoiding it, Major Scott ordered the helmsman to go straight through. This smashed a lot of crockery and damaged the cover, which required extensive repairs. On the return journey across the Atlantic the ship ran into a rainstorm and water poured into the accommodation area, putting both heating and cooking equipment out of commission. Fortunately the passengers and crew had been issued with fleece-lined flying suits. This was the twelfth and fastest crossing of the Atlantic by an airship. R 100 was then manhandled into a shed and never flew again.

In the meantime work was proceeding on the lengthening of R 101. In addition to this major refit, most of the outer cover had been replaced. The original material had been doped before it was fitted, and in 1929 was found to be in poor condition. This was thought to be due to the pre-doping, so the new cover was doped *in situ*. Two parts of the cover that had previously been treated in this way (cross-hatched in Fig. 5.7) were not replaced. Then on 24 September, just before the ship was due to be handed over to the flying staff, it was found that rubber solution had been

5.9 The lounge of the R 100. The person descending the stairway is the author Nevil Shute Norway.

used to attach patches and reinforcing strips to the old cover aft of the nose. Rubber solution applied over dope rots the cloth, but it was now too late for any replacement. So further reinforcing strips were stuck over the affected areas with red dope, which is a compatible adhesive. The ship was passed by the inspectors on 27 September, but could not be brought out of the shed until 1 October because of bad weather. Since the scheduled departure date for India was 4 October, the duration of the test flight was reduced to 24 hours from 48 hours. This flight took place during the night of 1–2 October, in clear weather with no turbulence. The flight instructions for the test required that the airship should run at full power from all five engines for at least five minutes. In the event an oil cooler failed on one of the engines and could not be replaced. Therefore the full power test was never done.

The disaster

Departure time was set for 18.30 hours Greenwich Mean Time on Saturday, 4 October, so that the ship would arrive at Ismalia, Egypt, in the early evening when, in the relatively cool conditions, there would be maximum lift. The weather forecast issued that afternoon predicted moderate wind over northern France, with light winds and clear weather over southern France and the Mediterranean. The conditions appeared to be promising; however, from the time the ship slipped its moorings they started to deteriorate. At first she was flown on a circular course such that it would have been possible to return to the mast (Fig. 5.10). Then, after about three quarters of an hour of flying, it was finally decided to proceed south *en route* to Egypt.

By this time it was raining and there was a strong south-westerly wind, with a cloud-base at 1500 ft. The cruising speed was 62 mph and height 1200 ft. Owing to the head wind component, however, the ground speed was just below 30 mph. Half an hour before midnight the airship crossed

5.10 The airship R 101 leaving the mast at Cardington.

the French coast and rose to 1500 ft to clear the higher ground. The weather was getting worse, with heavy rain, turbulence and gusts of up to 50 mph.

At 2 am on 5 October the morning watch took over (airships adhered to naval traditions). The second officer, Maurice Steff, was in command in the control car with chief coxswain George Hunt. At 2.07, after the ship had passed over Beauvais, Hunt came up into the chart room from the control car and told a member of the crew to go forward and release ballast. Then he called out 'we're down lads' (the chief electrician, Arthur Disley, who was in the switch room, heard this) and ran off, no doubt to warn the crew. At the same moment the ship lurched into a nose-down attitude and dived. After about a minute she came back on an even keel, and a telegraph message was sent from the control car to reduce speed. Then after a few seconds the airship dived again and hit the ground just about ten minutes after the start of the first dive. Within seconds a fierce fire broke out, consuming everything except the broken metal skeleton.

Harry L Leach, a foreman engineer, had gone to the smoking room (which was fire insulated) for a cigarette before turning in. When the ship levelled out he had time to pick himself up and replace some glasses and a soda-water siphon on the table. Then came the crash; the door of the smoking room flew open and he could see the flames surrounding the control car. Then the upper passenger deck collapsed but was supported at a height of about three feet by the room's furniture. At the same time a bulkhead fell out and he was able to crawl into the hull and thence on to the ground.

In number 5 after engine car Joe Binks was just taking over (a few minutes late) from engineer A V Bell. When the ship went into its second dive the engine telegraph pointer moved to 'slow', and Bell throttled down accordingly. Then came the crash, explosions and fire burning all around. The bottom of the car had been damaged and flames were creeping in, getting closer to the petrol tank used for the starter motor. Then suddenly a deluge of water came down – the fire had released some of the water ballast – and the flames around the car were extinguished. Binks and Ball were able to get out, pick their way across the field in pouring rain, and make contact with Leech and some of the other eight survivors (two of whom died later). Altogether 47 people were killed, including the Secretary of State for Air.

The fire came from two sources. Numbers of calcium flares were suspended around the control car. These were dropped into water to check heading and ground speed, and some had been used for this purpose during the channel crossing. They also ignited, disastrously, when the control car hit wet scrub. The other source was, of course, hydrogen,

5.11 The remains of R 101, Bois de Coutumes, Beauvais, 5 October 1930.

which could have been ignited either by an electric spark or by the hot exhausts of the forward engines.

The cause of the tragedy cannot be known with certainty. However, it is clear from survivors' evidence that the control car had been warned of a potential catastrophe. Joe Binks, as he came down from the crew quarters to the number 5 engine car, saw Michael Rope, assistant chief designer at the Royal Aircraft Works, making his way forward, apparently checking on the condition of the ship. It seems likely that Rope found a potentially fatal defect such as a major tear in the cover combined with deflation of the forward gasbags, and relayed this information to the control car by means of one of the speaking tubes that ran along the length of the ship. Consistent with this supposition is the known weakness of the cover in the nose sections, and the fact that the forward speed fell after the ship passed over Beauvais.

Faults and failures

This tragic, and rather sorry, story has been included here as an example of the potentially disastrous result of mixing politics and engineering. The Royal Aircraft Works was at one and the same time a contractor and the client; it was competing with a commercial organisation, Vickers, but at the same time was setting the rules. Not surprisingly, when it produced an inferior ship (as a glance at Table 5.2 will show) it disqualified the competitor and decided that its product was the only one suitable for the prestigious India route.

Subsequently, the whole programme was dictated by political considerations. The refit was scheduled to fit in with the date of a government conference. Because of this timing it was impossible to replace suspect parts of the cover, and in order to fit in with Lord Thomson's attendance at the conference, the tests required by the original specification were cancelled and even the substitute less onerous test was not properly completed.

Of course, even if all the engineering work and testing had been carried out satisfactorily, it is virtually certain that one or both of the airships would have come to grief within a few years. Most of the other big airships crashed, simply because they were only suitable for fair-weather operations, and by 1940 the few survivors had been grounded. Nevertheless the immediate fate of the R 101 and its distinguished passengers was the result of political pressure, which caused engineers not to complete work and tests which were essential for safety. And they were able to do this because, as a government agency, they could change the rules to suit political requirements.

Heavier-than-air machines

The fact that so many aircraft accidents are ascribed to pilot error suggests that navigational aids and similar devices should be high on the list of air safety programmes. So this is an important subject, and it will be covered separately in a later section. Firstly, however, it is necessary to look at the development of aeroplanes in general.[4]

One of the pioneers of manned flight was Sir George Cayley. Cayley established the basic design and layout of an aircraft and the essential requirements for stability. In 1853 he built a glider and this made a brief but successful flight, piloted by his coachman. In Germany Otto Lilienthal designed and constructed numbers of gliders, and made over 2000 flights before being killed in a crash in 1896. It was clear at that time, however, that no contemporary engine had the power-to-weight ratio which would make powered flight possible. The situation changed with the advent of the internal combustion engine. The Wright brothers made their first flight in December 1903 (Fig. 5.12). Their success was somewhat against the odds because the engine (which they made themselves) had a low power-to-weight ratio even for that time, and there was no tail or rudder.

There was more general interest in aeronautics in France than in America in those years and in January 1908 Henri Farman flew his own design of aircraft. In the summer of that year Wilbur Wright gave demonstrations in France of his pusher-type plane, creating great interest.

5.12 The Wright Flier: first to achieve controlled powered flight on 17 December 1903.

Nearly all subsequent developments, however, followed Cayley's principles, using either tractor or pusher propellers. Bleriot pioneered a tractor-type monoplane, and in July 1909 flew it across that stretch of water known on one side as the English Channel and on the other as *la Manche*. This was a dramatic achievement, and made Bleriot one of the immortals of flying.

The same year saw the introduction of the French Gnome engine. This was a rotary machine; that is to say, the crankshaft was stationary, and the air-cooled cylinders together with the propeller rotated around it. The Gnome was very successful and was used in many of the military aircraft produced during World War I.

The war provided a considerable impetus to aircraft construction. It has been estimated that in 1914 there were about 5000 aircraft worldwide, whilst in 1918 this number had risen to 200 000. They were used for observation and as bombers, and light, fast fighter aircraft were produced to shoot down the bombers. The bombers were in many cases adapted for use as passenger planes after the war; indeed, custom-built civil aircraft did not appear until the 1920s and even then their design was greatly influenced by their military predecessors.

Before 1914 air travel was uncomfortable and unreliable. Only a few operators were brave enough to try to provide scheduled flights, and these were usually for mail. There were aircraft, however, on the London–Paris route. The hazard for passengers was not so much that of being killed in a crash, but landing in a muddly field far from their destination.

After the war aircraft, and particularly the engines, became much more reliable, and regular passenger services became a practical possibility. In the summer of 1919 charter flights were operating between London and Paris, and (in Germany) between Dessau and Weimar. Continental European airlines were subsidised by governments, and this led eventually to the emergence of national airlines such as Lufthansa, Sabeana, KLM and Air France. In the UK this was not the case at first but in 1921 the operators refused to fly unless a subsidy was provided, this led to the formation of Imperial Airways in 1924. In the USA there was no such immediate post-1918 development; aircraft were used primarily for mail. The railroads offered a long-distance service which was safer, more comfortable, more reliable and cheaper than air travel, so initially there was not much incentive for airlines. However, in 1929 the Kelly Mail Act allowed the carriage of mail by private operators on scheduled services. Growth was then swift; by 1930 there were 40 small airlines carrying an annual total of 160 000 passengers. The big carriers such as American Airways, Pan American, Delta and TWA came into being during this period. Competition between airlines placed fresh demands on the aircraft manufacturers, and helped to develop the design and construction of larger and safer airliners.

The pre-jet age

Early aircraft were made of wood, wire and fabric (Fig. 5.13). The flight surfaces were formed by stretching cloth over wood and wire frames, and then doping the cloth. These covers did not constitute a major weakness as they did on airships, but occasionally they would disintegrate with disastrous consequences. Metal was, of course, more reliable, but although the first all-metal airplane, the Junkers F 13, was made in 1919, this type of construction did not become the norm until the late 1930s. In between these dates all sorts of materials were used. The original wood frame was in some cases replaced by high tensile steel tubing such as AISI 4340, joined together by oxyacetylene or arc welding. Eventually the frame structure was replaced by a stressed skin with stiffeners, as in the case of the de Havilland Comet described in Chapter 2.

Biplanes predominated in the early days because for any given weight it is easier to design and build than a monoplane. By the late 1930s, however, most civil aircraft were multi-engined, stressed skin all-metal monoplanes, and this general type of design has predominated ever since. So far as material is concerned, an exception was the de Havilland DH 91, the fuselage of which was made of a plywood/balsa/plywood sandwich. The wartime 'Mosquito' night fighter was also made of wood, but these two are probably the last examples of non-metal aircraft.

5.13 The Vickers Vulcan passenger aircraft. It cruised at 90 mph and had a range of 360 miles, carrying eight passengers. The fuselage was made of plywood.

The interwar years were the days of the flying boats. The advantages of these machines was that they required no landing strip or airport, and they had a long range. Imperial Airways operated flying boats on its eastern routes, and just prior to World War II started a transatlantic service, refuelling the seaplane in the air from a tanker aircraft. Dornier operated mail flights across the South Atlantic with Weil flying boats. The boats landed astern of a depot ship and were hoisted aboard and refuelled. Then they were launched by catapult over the bows to complete the last leg of the journey. Other flying boats operated in a more sedate manner, and provided relatively luxurious conditions.

World War II had remarkably little effect on aircraft design; civilian craft were adapted for military use rather than the other way around. It did, however, accelerate the production of a viable gas turbine engine. The possibility of such a machine had been known since the early years of the century, but its realisation had to await the development of alloys (the nickel-base superalloys) capable of resisting the severe conditions to which the blades and disc were exposed. A Heinkel HE 178 aircraft powered by a gas turbine flew in August 1939. In Britain Frank Whittle had run a prototype in 1937 and then, as war approached, received massive government backing. In fact, neither side made much use of gas turbine engines during the war, and it was left to civil aircraft designers to employ them on a large scale.

In the late 1930s and in the early post-1945 period American producers dominated the field of long-distance aircraft. First was the Douglas DC3 (Fig. 5.14), which by 1939 was carrying 90% of airline passengers. The DC3 was a twin-engined all-metal monoplane which, in terms of numbers, was one of the most successful aircraft ever built. The later development, the DC4, had important innovations; the cabin was pressurised for high altitude flight, and the wing structure incorporated multiple spars such that if a fatigue crack developed, enough load-bearing members would remain to prevent a catastrophic failure. The DC4 could carry 60 passengers, twice the capacity of the DC3.

Thus, at the end of the war, there were a number of aircraft adaptable to civil use, including the DC4, the Lockheed Constellation and the Boeing Stratocruiser, which were designed for long-distance operations, flying at 200 mph or more, and high enough to be above the weather. Safety, reliability and comfort had been enormously improved in a remarkably short period.

Jet age

The jet age was inaugurated by the setting up of regular transatlantic services in the late 1950s, with the Comet IV and Boeing 707. Both speed

5.14 The Douglas DC3, one of the most successful aircraft ever built.

and range increased dramatically. The cruising speed of the Boeing 707 is about 600 mph as compared with 315 mph for the DC6. The Comet and the Boeing 707 had turbo-jet engines, in which thrust is obtained by ejecting gas from the rear of the engine. Numbers of aircraft in the early years of the jet age, however, used the gas turbine engine to rotate a propeller. This variant did not survive very long, however.

Comet IV and the original Boeing 707 were relatively small planes; the original Comet 1A carried 44 passengers. Boeing jets grew fairly rapidly; the 707-320C took 219 passengers. Then the 747, which came into operation in the early 1970s could take nearly 500, the Lockheed Tristar 400 and the European Airbus 375. These large numbers made for uncomfortable journeys. The optimum was probably reached with the Vickers VC10, which had rear-mounted turbo-jet engines and flew quietly and smoothly with a complement of not more than 150.

Jet engines provide more than just increased speed and cruising height. Not only does greater power mean inherently greater safety in take-off, for example, but there are also other spin-offs. One of the most serious risks that aircraft face in cold weather is icing. Ice may build up on the loading edges of wings and tail plane, adding weight and making the plane difficult to control. Early de-icing devices included hollow rubber mouldings that were pulsated with compressed air in order to break off the ice, but these could be overwhelmed if conditions were really bad. Jet engines provided a virtually unlimited supply of hot air that could be fed to the leading edges so that the ice did not adhere.

Finally, there is supersonic Concorde (Fig. 5.15), which first flew in 1969, has a cruising speed of 1354 mph and a passenger capacity of just over 100. The sonic boom is a severe disadvantage, such that the aircraft can only reach its full cruising speed over the sea or over uninhabited country. Operating costs are high, and although a more economic version is possible, it is doubtful whether this will be built, at least in the near future. Although this looks like a dead end, Concorde pioneered some sophisticated techniques such as full electrical controls which have had a positive influence on aircraft safety; and it is one of the most beautiful aircraft ever designed.

Throughout this history of rapid development and change, the materials of construction have, on the whole, altered very little. The first all-metal aeroplane was made in 1919 using Duralumin, and most aircraft flying today are built of the same or similar age hardening aluminium alloys. A recent advance has been the introduction of aluminium–lithium alloys, which are lighter. Some supersonic military aircraft have a titanium skin, but this metal is still too costly for civil planes. Carbon fibre reinforced plastic has found a limited use, for

5.15 Concorde in flight.

example, for the rudders of the Airbus 300 and 600, and glass-reinforced plastic may be employed for lightly stressed parts. Large changes in materials and basic format are unlikely in the near future. More development is possible in electronic control, as discussed below.

Control systems

These systems fall into one of several categories. The first comprises the means by which cockpit controls activate the moving parts such as ailerons and rudder. Secondly, there are automatic pilots. Then there are instrument landing techniques: aids to landing in bad weather. Air traffic control is a fourth area and finally there are systems that inspect for, and report, faults in other systems – very much a product of the electronic age. The techniques that fall into such categories are all intended to improve safety.

Autopilots

It seems appropriate to consider autopilots, first because they came first; Almer Sperry designed his first autopilot in 1909, at a time when there were very few human pilots. Five years later in 1914 a Curtis seaplane flew automatically across the river Seine, guided by Sperry's son Laurence. The autopilot used a gyroscope to detect roll and pitch, and operated hydraulic servo controls to maintain stable flight. Similar systems remained in use until the 1930s, when full electrical operation came into use. Then during the 1940s, electronic devices were introduced. In 1947 a Douglas C-54 flew from Newfoundland to England with the autopilot in charge throughout the flight including takeoff and landing. Developments have continued at an accelerated pace with the introduction of increasingly powerful computers.

An autopilot may be used in all phases of flight; in climbing, in level flight and in the descent, although the final part of the approach has usually been controlled by signals from the airport, as will be described below. Automatic control has recently been used to counteract the effects of turbulence.

Instrument landing systems

The earliest system relied on quick thinking by the pilot. With some guidance from the control tower, the pilot would dive into the low fog or mist, and if, when there was some visibility, the runway was straight ahead, the plane was landed. If not, the pilot overshot, climbed, turned and tried again. Not a very relaxing procedure, particularly if it had to be repeated more than once.

Since 1950 major airports have been required to install instrument landing systems. The arrangements are illustrated diagrammatically in Fig. 5.16. There are two radio transmitters; one (the localiser) emits in a vertical plane which runs along the centreline of the runway; the other emits along the glide path. There are also beams that are directed vertically to show the distance from the start of the runway. The first two signals activate an instrument in the cockpit, which operates as shown in the inset to Fig. 5.16. There are two pointers; when they are respectively vertical and horizontal the aircraft has the correct heading and is flying directly down the glide path. Deviations are indicated when the pointers are in other positions, as shown.

In his approach, the pilot makes a turn which is continued until the localiser beam is picked up. The aircraft is then lined up with the beam and descends until it picks up the glide path indicator. The glide path is held until there is sufficient visibility for a manual landing. At this point

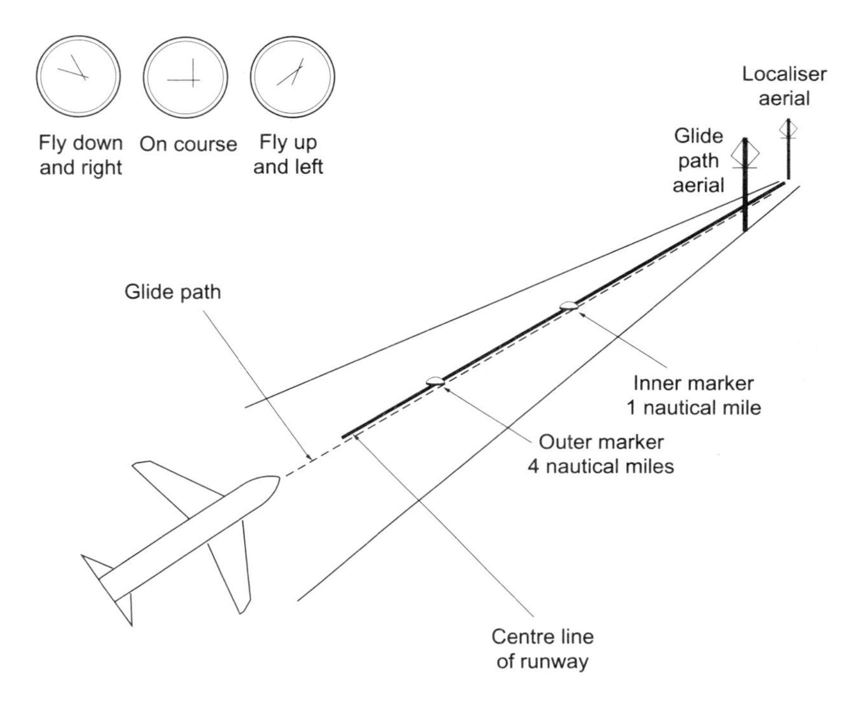

5.16 Instrument landing system.

the pilot decides whether or not a landing is practicable; if not, he overshoots.

A completely blind landing was, it is claimed, first achieved at Farnborough in 1945, during the wartime blackout. However, it was not until the 1960s that there was sufficient confidence to use such techniques on passenger-carrying flights. One of the first aircraft to operate in this way was the Vickers Trident, a three-engined short-range jet used by British European Airways. The Caravelle, a very successful French rear engined jet, was also licensed to carry out blind landings.

Flight controls

In the early days pilots operated their control surfaces such as ailerons, rudders and flaps directly, by means of wires and rods. Later, servomotors were added to do the manual work. However, starting with Concorde and in most of the recently built large jets, movements of control levers in the cockpit are translated into electrical impulses and these in turn operate the flaps, etc. This is known as the fly-by-wire system, and the wires are shortly to be thrown out in favour of optical transmission, which is lighter.

Such a system facilitates the type of computer control mentioned earlier; it is lighter than mechanical control, and it is cheaper to install and maintain. On the other hand electrical, and particularly electronic, systems are vulnerable to damage in electric storms. So Concorde, for example, has a backup control system. It is hard to see how safety can be assured without such backup.

For civil aircraft it is generally accepted that the risk of catastrophic failure must be less than 10^{-7} per hour. Since the failure probability of the more sensitive electronic or electrical systems in a fly-by-wire aircraft can be up to 10^{-4} per hour, this means that in the case of automatic flight control, for example, it is necessary to duplicate, triplicate or quadruplicate circuits. There may be a considerable weight and cost penalty for providing such multiple systems and in some cases it is possible to use detectors that are capable of finding the fault and reconfiguring the system sufficiently quickly (typically this must be done within about 0.2 seconds) to avoid catastrophe. The technology is sophisticated and appears to be successful; Boeing put less than 2% of total aircraft losses in the category 'instruments and electricals'.

Air traffic control

The first international agreement on the control of aircraft movements was achieved at the Versailles Peace Conference in 1919, and a high level of international co-operation is still the rule. The International Civil Aeronautical Organisation now looks after and approves air traffic control systems worldwide.

The first air traffic controllers were men with flags. Waving the flags cleared the aircraft for takeoff. In the late 1920s and 1930s custom-built airports began to appear, one such being Croydon airport, south of London. A feature of these airports was the control tower, from which aircraft movements were co-ordinated. The first radio-equipped tower was at Cleveland Municipal Airport in 1930. In the USA the principal airlines established a control system based on Newark, Chicago and Cleveland, and in 1935 the US government took over the operation and increased the number of centres quite rapidly. In the early days control was by direct communication between tower and pilot. Aircraft were cleared for landing or takeoff by verbal instruction.

A major step forward came with the installation of radar. This enabled the controller to see the aircraft on his screen, but the identity and altitude were obtained from the pilot. Then methods of identifying the aircraft and measuring its altitude were developed, such that much of the information is processed by computer and displayed on a screen. In the latest technique, due for installation in Europe and the USA during 1995–98, the operation will be automated, and clearance will be given automatically by the computers. However, air traffic controllers retain the responsibility for separation of aircraft, which is 8–16 km horizontally and 300 m vertically.

After obtaining clearance for departure from the airport tower, an aircraft operating over land passes through a series of geographical sectors, each of which is individually controlled. The sector controllers clear the aircraft to enter their airspace and eventually hand the aircraft over to the neighbouring controller. These operations are guided by display screens which show the position and movement of each aircraft, as determined by radar. If at any point in the chain a controller cannot accept an aircraft, it is stacked in a holding pattern which maintains the specified vertical and horizontal separation. No system is perfect, of course, but the technological improvements in this field have greatly reduced the risk of mid-air collision.

Passenger safety

In view of the phenomenal rate at which aeronautics has progressed since the beginning of the twentieth century, it is not surprising that safety has likewise improved. Nearly all the technological change has been in a positive direction so far as safety is concerned: the rapid increase in engine power, the increased strength of the airframe due to all-metal construction, automation of controls, the development of blind landing methods and improved air traffic control have all made contributions to a reduction in fatality rates of (in recent years) about 5% annually. No doubt this improvement will continue. It was noted in Chapter 1 that the introduction of new aircraft types usually results in a short-term increase in loss rates. The phase of rapid and sometimes revolutionary change in design of both airframe and power units which produced these temporarily unsafe conditions has probably come to an end, and the future is more likely to see steadier and less radical changes. The situation should in itself promote safety.

Shipping

The Lloyd's Register data quoted in Chapter 1 showed a steady fall in the percentage loss of shipping extending from the 1890s to the 1990s. The fatality risk for seamen has fallen in a similar way. In 1900 the annual death rate was 0.14%, and by 1950 it had fallen to 0.03%, corresponding to an annual decrement of 3%. Recently casualty rates have fallen in a similar way. Such improvements have resulted very largely from improved technology, although legislation and the regulations of the International Maritime Organisation have no doubt played some part, and more recently better rescue services have helped.

Technological change

The nineteenth century saw changes in marine technology that were more profound and far-reaching than any that had gone before. The change from sail to steam, and from wood to iron and then to steel, occurred more or less at the same time, beginning about 1820 and being more or less complete by the beginning of the twentieth century.

The American ship *Savannah* is credited with the first steamship crossing of the Atlantic in 1819. However, the *Savannah*, in common with other early vessels of its type, was really a wooden sailing vessel with an auxiliary engine driving paddle wheels. The paddle wheels could be detached and stowed on deck, and this is where they stayed for all but eight hours of the 21-day voyage. In practice, most of the early steamships were packet boats, providing a regular passenger service on coastal or cross-channel routes. They carried sails which could be used to save coal in a favourable wind or when the engine broke down. The steam engines themselves were very large, with pistons having a stroke of up to 6 ft and diameters as large as 7 ft. At the time boiler shells were made of low-strength iron plates, and the maximum steam pressure available was about 5 psi, so to obtain sufficient power a large diameter cylinder was required.

Paddle steamers were not comfortable ships in high seas. It was necessary to locate the paddle shaft well above the water line so the centre of gravity of the machinery was high, consequently the ships rolled badly. When they rolled, one paddle came out of the water and the other ploughed in, causing a corkscrew motion. Nevertheless, unlike sailing ships they were independent of the wind, and could maintain regular schedules with much shorter journey time. The steamer companies thus got the mail contracts, and this helped to finance expensive development (as happened with aircraft nearly a hundred years later). Two lines in particular benefited in this way: Cunard, which had the main share of the transatlantic traffic, and P & O (originally the Peninsular and Oriental Steam Navigation Company), which pioneered the Eastern routes, eventually to India. Figure 5.17 shows the *William Fawcett*, considered to be the first steam paddle ship operated by P & O for regular passenger and mail services, and Fig. 5.18 pictures one of their early screw-driven ships.

The screw propeller was invented in 1838 but it was a long time before

5.17 A model of the 206-ton *William Fawcett* built in 1828, intended for the P&O service between the British Isles and Spain and Portugal.

5.18 The 3174 ton *Hong Kong* propeller steamship built for P&O in 1889.

it was universally adopted. Cunard took delivery of their first propeller-driven ship in 1862, partly because Samuel Cunard, who founded the company, thought the old ways were best. Isambard Kingdom Brunel took a different view and his *Great Britain*, launched in 1843, was the first large ship to use a propeller. It was also one of the early iron ships. Iron provided the rigidity that was necessary for the propeller and its shaft.

Not all Brunel's innovations were successful. A major disadvantage of wood as a structural material was that the greatest length of ship for which it could be used was about 300 ft. Brunel's iron ship *Great Eastern*, launched in 1859, was over twice as long with a displacement of over 18 000 tons. It was by far the largest ship afloat at the time, and it was equipped with both screw propellers and paddle wheels. However, on her first crossing of the Atlantic, high seas broke off the paddle wheels and smashed the rudder. During that period Atlantic liners used to carry a cow to provide fresh milk for the passengers. Unfortunately the broken paddle demolished the cow-house and precipitated the cow through a skylight on to the passengers in the saloon below. Subsequently the *Great Eastern* failed to attract passengers and finished up laying the transatlantic telephone cable.

The first iron ships were constructed about 1840, and by 1890 very few wooden ships were being made (Fig. 5.19). Towards the end of the nineteenth century steel became cheaper than iron so shipbuilders and designers were able to take advantage of its higher strength. Steel also enabled boiler manufacturers, who had already developed iron boilers to a considerable extent, to increase steam pressure still further. Higher pressures made compound engines possible, in which the exhaust from a small high-pressure cylinder passed to a larger low-pressure cylinder. This principle was extended to triple expansion and quadruple expansion

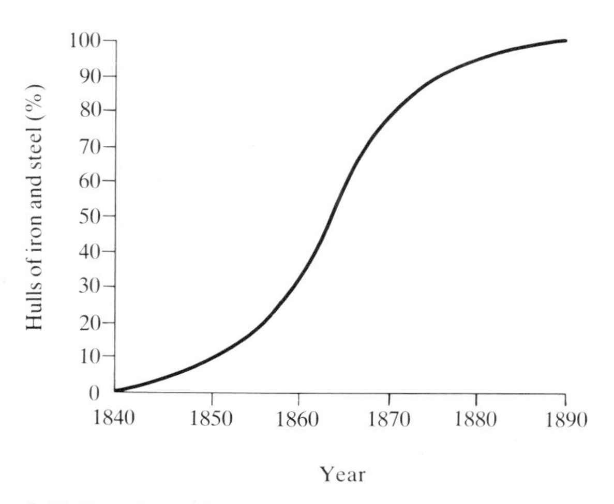

5.19 Ships built of iron or steel as a percentage of the total 1840–90.

engines, which provided greater power and were more economical. The ultimate development of these reciprocating steam engines was a quadruple expansion type working with superheated steam at 400 psi, incorporating reheat.

The steam turbine was developed by the end of the nineteenth century and became the preferred type of engine for warships and large passenger ships. After World War II they were also used in cargo ships, container ships and tankers. Diesel engines also became competitive for ship propulsion between the world wars and have become increasingly dominant in recent years.

The first steamers had a displacement of a few hundred tons, and were very small ships indeed by current standards. The size increased quite rapidly, particularly in the case of transatlantic liners, which set the pace. The first of these, the Cunard ship *Britannia* was 207 ft long and 1145 tons. It was a square-rigged wooden paddle steamer, and very uncomfortable. Charles Dickens took passage to Boston on her in 1842. By the 1880s the sails had almost gone and the tonnage had gone up typically to about 7000. About this time the notion that a passenger liner could be a floating hotel began to take shape, and the next generation, including two ill-fated ships, the *Lusitania* and the *Titanic*, had very luxurious first-class accommodation. The *Lusitania* displaced 31 550 tons and the *Titanic* 46 383 tons. The final phase for the big transatlantic liners, before their trade was taken over by the airlines, came with the launching of the *Queen Mary* in 1934, at over 80 000 tons displacement. Liners still operate on pleasure cruises, but they have not continued to grow in size. The *Oreana*, launched in 1995, displaces 69 153 tons.

Size is clearly an advantage in the battle against wind and weather, but it is no guarantee of safety, as the loss of the *Titanic* demonstrated. It might be thought that at least it would ensure a smooth crossing, but this too was not always the case. The Atlantic weather in 1936, the year of the *Queen Mary*'s maiden voyage from Liverpool to New York, was the worst that anyone could remember. Unfortunately the new ship rolled badly. At the height of one storm an upright piano in the tourist class lounge tore itself loose and crashed around, smashing furniture and tearing off the panelling. A number of passengers were injured, and when the ship arrived in New York, to a tumultuous welcome, there was a line of ambulances waiting, parked discreetly out of sight. The problem took a year to rectify.

Cargo vessels went through a similar cycle of growth, which culminated in the construction of great supertankers, followed by stabilisation at a more modest size. One of the vital factors in the growth of the cargo fleet – numerically rather than in size – was the establishment of a network of telegraphic communications in the industrialised countries. The first successful submarine telegraph cable was laid between Britain and France in 1851, and a transatlantic link followed in 1858. This system made possible the operation of tramp ships that, having discharged a cargo, received instructions as to where next to proceed. The tramp steamer was a relatively small general-purpose ship that dominated the scene until after World War II, when specialised vessels started to appear. There were oil tankers already, but bulk carriers for such cargoes as iron ore and grain were developed. Then in the 1960s the container revolutionised the transit of cargo. Goods were stowed into standardised containers at warehouses remote from the ports and loaded mechanically on specially adapted ships. This eliminated the traditional dock labour and greatly reduced pilfering, which had been an ancient tradition at ports and a constant source of loss. Special ships were also made for transporting liquefied natural gas and other refrigerated cargoes.

There is no doubt that standards of both design and construction have improved substantially in the post-1945 period. Failures such as the Liberty ships and the *Sea Gem* cast a long shadow, and materials have been radically improved, as described earlier. Design has, with the assistance of the computer, become more exact and sophisticated. Altogether it is becoming less and less likely that failures will occur because of gross deficiencies in material, design or construction. It is not, of course, possible to provide for the worst excesses of storm and tempest, nor is it possible to eliminate human error. But the navigational aids described below are improving progressively, in such a way as to minimise these two hazards.

Navigation

In the nineteenth century the navigation of a ship relied on three essential aids: an accurate chart, the ability to determine position and the measurement of the ship's speed and an estimation of its drift due to wind and current. Latitude was measured by making astronomical observations with a sextant. Longitude was obtained by noting the time at which the sun had reached its zenith. Every four minutes after 12 o'clock noon represented one degree east of the Greenwich meridian, and vice versa, so that an accurate chronometer was required. Speed was obtained by throwing a plank over the stern of the ship. Attached to the plank was a cord which was knotted at regular intervals, and the seaman counted the number of knots that passed through his hand during a period of 30 seconds: this gave the speed, in knots, naturally. Having determined latitude and longitude, the navigator was able to use his chart in order to measure the course to be set, and this course was maintained by the helmsman by means of a magnetic compass.

By the year 1900 the required skills were so developed that a good navigator could, in favourable weather, determine a ship's position with reasonable accuracy. Fair weather was not always to hand, though, and the advent of wireless was a major step in resolving the remaining uncertainties.

Wireless (originally wireless telegraphy, now radio) became a practical aid to marine navigation after Marconi made the first transatlantic transmission from Cornwall to Massachusetts. Wireless quickly became standard equipment on large vessels. It enabled them to maintain contact with shore and with other vessels, and to send out distress calls when necessary. Time signals were sent out daily from Greenwich, enabling navigators to correct their chronometers. In the original Marconi system, the radio signal was generated by an electric spark and comprised a spectrum of wavelengths; receivers converted these signals into an on–off electric current. Voice transmission was not possible and all communication was made using the Morse code. The practical effect, however, was dramatic, and great public interest was aroused; for example when the American ship *Republic* collided with the Italian *Florida* in 1909, wireless signals from the *Republic* brought help within half an hour and the entire complement of both ships, amounting to 1700 passengers, was saved. Then in the same year Dr Crippen, having murdered his wife in London, was escaping with his mistress to the USA when, as a result of a wireless message, he was arrested on board ship. And then there was the *Titanic* disaster. Few, if any, technological changes have had such a startling impact as radio communication.

Following the invention of the thermionic valve in 1904 and subsequent developments in electrical circuits, the transmission of speech became possible just before World War I. The internationally agreed distress call in speech was 'Mayday' which is an anglicisation of the French *m'aidez*.

Even before these developments, it was found possible to use a radio receiver to determine the direction from which a signal had come. Radio beacons had begun to supplement or replace lighthouses, and a ship could find its position without astronomical observations and in any weather other than a severe electrical storm. The sextant was put away for good.

In about 1925 echo-sounding devices came into use, replacing the ancient lead weight and line method. In their present manifestation, echo-sounding (also known as sonar) systems are fitted below the hull of the ship and emit a series of ultrasonic pulses. Electric currents measure the time of reflection of the pulses from the sea-bed and convert such measurements into distance. The result is displaced on a cathode-ray tube, or a piece of paper. This aid to navigation was used to detect submarines during World War I.

During the 1930s the possibility of using radio waves for echo-location began to be explored, particularly in Britain, Germany and the USA. The Germans had produced such a device in 1933, but they used too long a wavelength and the equipment was not very effective. In fact, useful results could only be obtained with a high frequency beacon operating in the range 300–3000 megahertz. The beam is generated by a device called a magnetron, and is fed to a rotating parabolic transmitter. Reflections are picked up by an antenna and the results displayed on a circular cathode-ray screen. Solid objects appear as a bright spot. Their distance from the centre of the screen represents the range, and the angular displacement from the vertical is the bearing. The main incentive for this development was the need to detect aircraft during World War II, but its value for the avoidance of collisions at sea will be obvious, and radar is now standard equipment for most seagoing vessels.

Computer technology and satellite communication have further improved the navigability of ships in more recent years.

All in all, radio has conferred tremendous benefits to shipping, and must have been a major factor in the improvement of safety that we have seen during the twentieth century. However, many of the other improvements that have been described here have also made a contribution, and there is little doubt that the future will see yet further improvements in marine safety.

Bulk carriers

Bulk carriers are vessels designed to carry very large quantities of cargo such as grain, coal or iron ore. In effect, they are rectangular boxes divided into compartments, with machinery and crew accommodation tacked on the stern, as shown diagrammatically in Fig. 5.20. Typically such vessels could be 1000 ft long, 100 ft high and 150 ft wide, and contain 30 000 tons of welded steelwork. Bulkheads are usually corrugated in form, and there is a complex system of internal stiffeners, attached to the steel plates by fillet welds. In many cases the material used has been Lloyd's Grade A

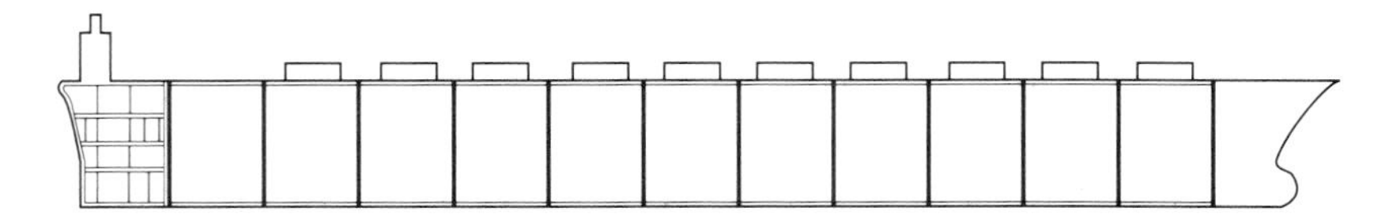

5.20 Diagrammatic cross-section of a bulk carrier (after Ref. 9).

steel, which is a general purpose carbon steel with no impact test requirements.

Numbers of these ships have been lost, and some have disappeared without trace. More important is the human loss, which has averaged about 150 lives per year. The *Derbyshire*, which was built in 1976, sank with all hands during a typhoon in 1980. The wreck was located on the bed of the Pacific Ocean in 1994 at a depth of 14 000 ft. This particular loss has caused much concern, because no radio distress call was received, suggesting the possibility of a sudden catastrophic failure.

There are two main causes of deterioration in bulk carriers. The first, as would be expected, is fatigue cracking associated with fillet-weld attachments to hull and deck plates. In some cases these have led to local brittle fractures, and in others to areas of plating being completely lost. The other problem is corrosion. This has mostly been caused by wet coal cargoes. Water reacts with sulphur in the coal to form acids, and these in turn corrode and thin the plating.

The steps required to mitigate these losses remain to be determined. Proposals include the use of impact-tested grades of steel for the plating, and developing adequate means of inspection for fatigue cracking. The problem has been reviewed by J Jubb.[5]

Offshore drilling and production

The percussion method has been used for drilling deep wells by the Chinese for over a thousand years, and came into use in Western Europe and America towards the end of the eighteenth century. This technique employed a metal bit shaped like a chisel to break up rock and bore through it. The tool was alternatively raised and then allowed to drop; for example in the Chinese method the bit was suspended by a rope from one end of a plank which pivoted like a sea-saw, and the drillers took turns to jump on the end of the plank, so giving the tool a reciprocating motion.

Percussion drilling was used mainly to extract salt, which often occurs as 'salt domes' lying below a cap of rock. When Colonel Drake drilled the first oil well in Pennsylvania in 1859 he employed a salt driller to do the work. By this time the steam engine had been harnessed to provide the up-and-down motion, as shown in Fig. 5.21. The tool was now suspended on a wire rope, and a derrick provided the means to hoist the bit and other devices into and out of the well. The system became known as cable tool drilling.

The rolling cutter rockbit was invented by Howard Hughes and came into use during the first decade of the twentieth century. This tool became more generally used in subsequent years but cable tool drilling continued

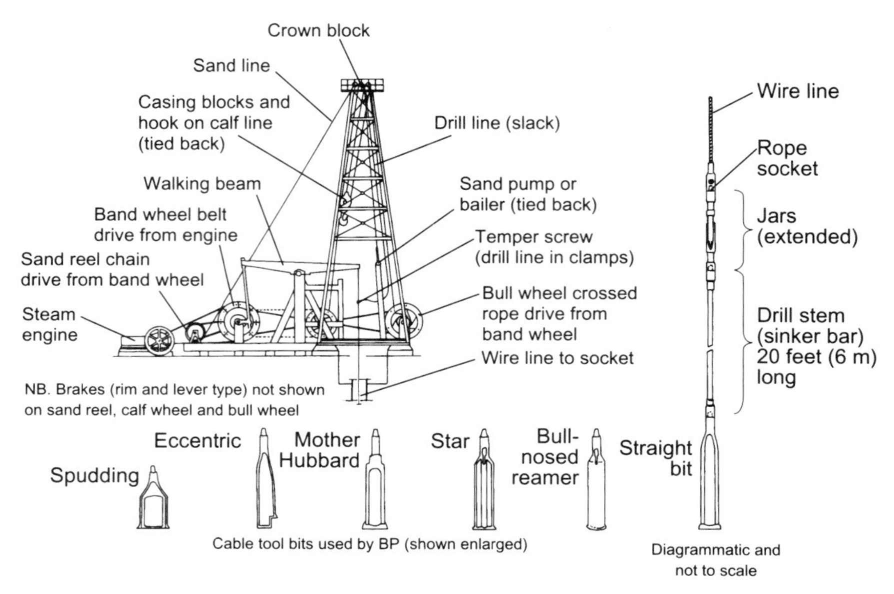

5.21 Cable-tool drilling rig.

to dominate the scene. Then the financial crisis of 1930 halted oil drilling completely for two years. When operations resumed, the competitive climate favoured rotary drilling and the cable tool fell out of use.

Rotary drilling introduced a radically new method of removing the spoil. Previously this had been taken out periodically by a bailer, which was essentially a tubular can with a self-closing bottom. In rotary drilling a circulating fluid was employed which flushed out the ground rock and conveyed it to the surface. In making a vertical hole, the drill is rotated by a tubular shaft known as the string. A liquid flows down inside the string and then up through the annular space between the string and the well casing. As well as carrying away spoil this fluid (called mud) cools the drill head and provides a hydrostatic head which helps to counter the pressure in the oil-bearing formation (down-hole pressure). The mud is circulated from a tank down and up the well, then over a shaker which separates the chippings, and back to the mud tank (Fig. 5.22).

Oil wells go very deep and the pressure is correspondingly very high. It is normal practice to fit a non-return valve in the string just above the drill head to prevent an uncontrolled upward flow. However, accidents may happen; the string may fracture or come unscrewed, or the drill may break unexpectedly into a high pressure region. Therefore the wellhead is fitted with means of countering such emergencies. Figure 5.23 illustrates a blowout preventer of the type commonly fitted to land-based wells and

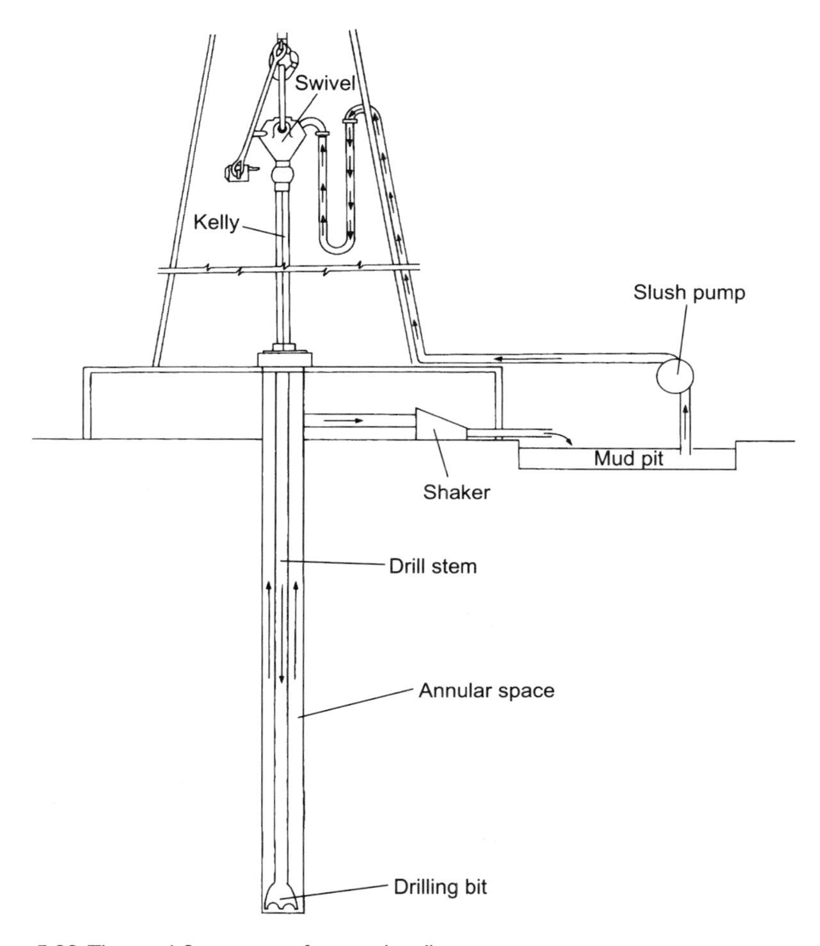

5.22 The mud flow system for an oil well.

some offshore wells. The top preventer has a large rubber element capable of sealing around any tool or around the drill string. The middle ram will close the annular space around the string, while the lower, blind ram consists of two flat-faced elements that meet in the centre and seal off the well completely. If a drill string is present, the tube is crushed. These closures may be individually and separately operated by the drilling crew according to the prevailing emergency. In less pressing circumstances an excess in down-hole pressure may be countered by increasing the density of the mud. This is usually accomplished by the addition of barytes (barium sulphate).

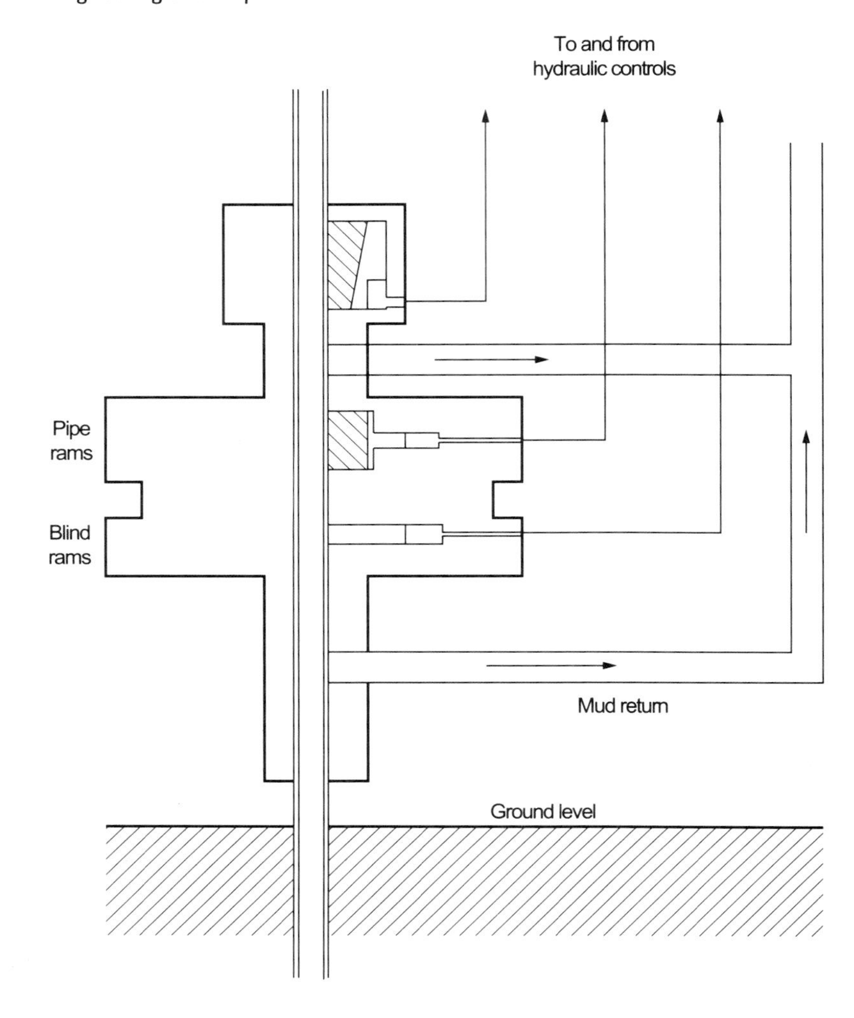

5.23 Well-head with blowout preventer: sectional diagram.

Oilfield development

There are three main stages in the establishment and exploitation of an oil or gas field. The first step is to carry out a seismic survey to identify rock structures that could be oil-bearing. Offshore this may be done by detonating a propane–oxygen mixture in a rubber container and analysing the reflections from the sub-sea strata. An alternative source is a high energy vibrator. Explosives are no longer used because of the damage caused to marine life.

When a likely area is found its boundaries are determined and then an

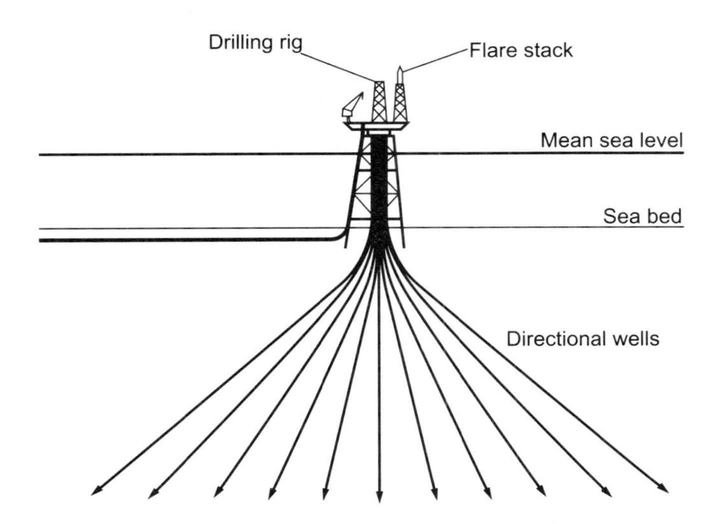

5.24 Development drilling from a fixed offshore platform.

exploratory vertical well is drilled somewhere close to the centre. This is the wildcat well, and is generally considered to be the most hazardous of the drilling operations, because the composition and pressure of any hydrocarbon that may be present is unknown. If results from the first well are positive, then other exploratory wells are drilled with the object of defining the extent of the field.

In the third phase an offshore field is developed by establishing fixed platforms and using these as a base to drill production wells. Sometimes there will be only a single vertical well; more often multiple wells are sunk.

In order to cover the area assigned to the platform, wells must be dug at an angle to the vertical (directional wells: Fig. 5.24). To do this, turbodrills are used. These drills are driven by a water turbine mounted behind the head, and the turbine itself is driven by mud flow down the string. Using this device it is possible to drill wells at angles of up to 45 ° from vertical. As these are completed the wells are tied in to the separation unit and then via an export riser to the sub-sea pipeline that takes the product onshore.

Blowouts

Blowouts are a particular hazard offshore because not only are they a danger to life and limb but, in the case of any oil blowout, they also generate an oil slick and much unfavourable publicity. The majority of blowouts, however, result from the sudden release of gas.

Figures gathered by the US Geological Survey indicate that for the period 1953–71 blowouts occurred in 0.02% of the wells that were drilled, and that this rate did not vary significantly.[6] A later report from the Minerals Management Service of the US Department of the Interior covering the period 1971–85 also indicated no significant change in the blowout rate. The WOAD data for 1970 to 1987 shows that blowout accidents that resulted in fatalities did not vary significantly. About one-third of the incidents related to fixed units (jackets) and presumably took place when drilling development wells. The risk is evidently not confined to exploratory drilling. Indeed, referring to Table 1.14, it will be seen that if supercatastrophes are excluded, blowout accounts for 36% of all fatalities on fixed units, as compared with 13% on mobile units.

Blowouts are an intractable problem. It would appear that blowout protectors, if properly maintained and operated by skilled people, are reliable and capable of controlling a well in an emergency. However, a rapid build-up of pressure or loss of circulating fluid, or simple human error can all lead to loss of control, and it is difficult to see how technological progress can have very much effect on this situation.

The development of offshore structures

The drilling of underwater oil wells started during the 1920s in Louisiana and in Lake Maracaibo, Venezuela. Piles were driven into the lake bed and these supported a relatively small structure and platform, on which were mounted the derrick and power units. Ancillary equipment was carried on a tender moored alongside. When the well was completed the derrick was removed, leaving the platform with its production well-head. A similar type of operation was conducted from jetties or piers off the Californian coast.

Increasing demand for oil and gas after World War II forced operating companies to consider extending their activities to the continental shelf. In 1945 President Truman signed a proclamation laying claim to all the mineral rights in the regions off the coast of the United States. This claim was confirmed by the Geneva Convention on the Continental Shelf in 1958. Meantime there has been continued development of offshore technology. In the 1940s the first self-contained offshore units were built. These were flat-bottomed barges which were towed out to the required position. By flooding buoyancy compartments they were made to settle on the sea-bed with the drilling platform above the waterline. Such operations were mainly confined to shallow water in the Mexican Gulf.

Platforms of this type are designated 'submersibles'. In recent years sub-sea vehicles or submarines have also been categorised as 'submersibles', and so there is scope for confusion. So far as this book is concerned submarines are submarines and submersibles are the type of offshore platform referred to above.

In the 1950s, to cope with the requirement for drilling at greater depths, submersibles were further developed by building the platform on columns supported by a submerged flooded pontoon, and such units have operated in up to 175 ft of water. Jack-up platforms appeared during the same period; these, as we have seen in the case of the *Sea Gem*, were adaptations of self-elevating platforms used for general engineering purposes. In the late 1950s two types of floating platform appeared; the semi-submersibles and the drill ship. Semi-submersibles, which float on submerged pontoons, are held in position by a system of anchors generally similar to that described for the *Alexander L Keilland*, and can drill to depths of 1500 ft. The drill ship, which can drill down to 6000 ft or more, appeared at about the same time. At first, this was an adapted merchant vessel, moored like a semi-submersible, but in the 1960s, purpose-built vessels with dynamic positioning equipment were introduced. Figure 5.25 illustrates these developments.

Submersibles, semi-submersibles, jack-up platforms and drill ships are all mobile units, and are normally used for drilling exploration wells. Development wells are drilled from fixed units, which may also carry primary separation equipment and gas or oil exporting pipelines. There are four main types: jackets, artificial islands, concrete structures and tension leg platforms. Jackets have been described elsewhere; Piper Alpha was typical of such platforms. Artificial islands are exactly what the name suggests: they are made by dredging sand or gravel from nearby and dumping it to make an island, from which drilling operations can be conducted more or less as on dry land. Artificial islands may have an advantage in shallow arctic waters where structures could be damaged by

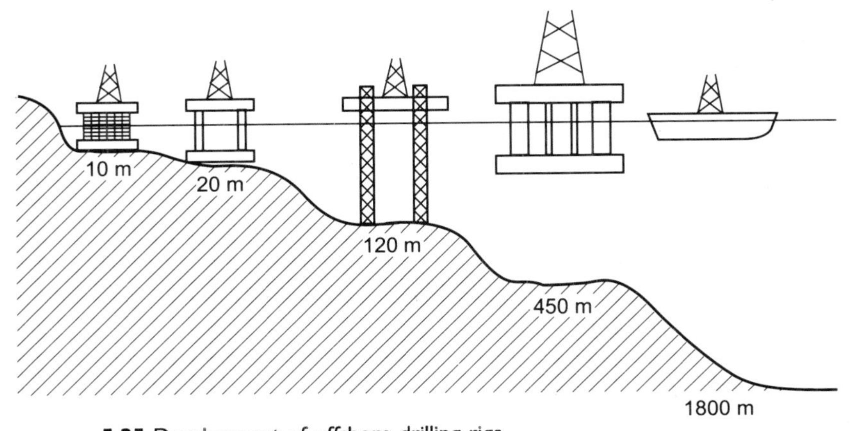

5.25 Development of offshore drilling rigs.

ice. Reinforced concrete is used for very large bases that rest by gravity on the sea-bed. A platform of normal metal construction is mounted on top of the concrete columns. Tension leg platforms resemble jackets, but are mounted on pontoons, like a semi-submersible, to give positive buoyancy. The whole system is tethered to the sea-bed by tension members. Tension leg platforms are intended for operating in waters that are too deep (over about 800 ft) for a conventional jacket and are coming into use as a means of extending operations into deeper waters. Figure 5.26 shows the Conoco Hutton tension leg platform which is operating in the North Sea.

The record is one of rapid technological change, in which completely new types of drilling platform have been developed. It is not surprising in these circumstances that the fatality rate has been high. During the period 1970–87 nearly 60% of the casualties occurred as the result of four major losses of mobile units. One of these, the *Alexander L Keilland*, was of a type that is unlikely to recur. The others were sunk in heavy seas, and whilst this risk will always remain, improved stability and better evacuation procedures could greatly reduce the fatality rates. One of the negative factors in the case of offshore operations is that the improvements in navigation and guidance systems that have had such a beneficial effect for aircraft and for normal shipping do not make much difference to the safety of mobile drilling units. Moreover, improvements in safety are to some degree countered by the need to operate at increasing depths. Nevertheless, experience in the 1980s had led to some useful developments. The very radical changes in the layout of jackets following the Piper Alpha disaster are described in Chapter 2; these represent a substantial contribution to safety. After the *Alexander L Keilland* accident and the capsizing of *Ocean Ranger* in 1982, the classification societies required that redundancy should be incorporated in the structure and that the platform should be capable of floating. Some of the more recent semi-submersibles have accommodated these requirements by using twin floaters (pontoons) on which are mounted braced rectangular columns supporting the platform.

Hydrocarbon processing

In 1784 Ami Argand, a Swiss distiller, invented a new type of oil lamp, which was to usher in great improvements in domestic comfort. In earlier times oil lamps produced a deal of smoke and smell, but very little light. Argand used a circular wick around which was mounted a glass chimney, such that air was drawn up on either side of the wick. This produced a bright clear flame, giving a light equal to 10 or 12 tallow candles. The same principle is used in oil lamps to this day. Unfortunately

5.26 The Conoco Hutton tension leg platform.

for Argand, his partner allowed the patent to lapse and within a year or two the design had been pirated and 'Argand Patent Lamps' were on sale in London.

Gas lighting and the means of producing illuminating gas from coal were developed in the 1790s by William Murdoch, and by 1802 had been used to light a factory in Birmingham. The streets of Baltimore were lit by coal gas as early as 1816. Domestic use developed more slowly, partly because of the need for a piped supply system, and was restricted to towns. In country areas (and in the nineteenth century a high proportion of the population lived in the country) illumination was, other than in a few big houses, by candles and oil lamps.

The oil used for the Argand lamp and its subsequent developments was rape oil (then known as colza oil), or alternatively whale oil. These liquids were too viscous to be drawn up by a wick, so various ingenious means were used to bring the oil close to the flame. In the Argand and other lamps it was fed by gravity, and in others by mechanical devices. Kerosene, known in Britain as paraffin, is non-viscous and can readily be drawn up a cotton wick. In the 1840s and 1850s it was obtained by distillation from coal and from oil shale. So when Colonel Drake struck oil in Pennsylvania in 1859 there was a large established market for lamp oil. It may be argued, therefore, that Ami Argand was the true father of the oil industry.

Initially kerosene was the only marketable produce to be obtained from crude oil; the remainder was discarded or burnt. However, during the last quarter of the nineteenth and first half of the twentieth century the gas industry found it expedient to supplement supplies of coal gas with semi-watergas, which is made by passing a mixture of air and steam through red-hot coke. This produces a mixture of hydrogen, carbon monoxide and nitrogen which burns well but has a lower calorific value than coal gas. To make up the difference, gas oil, which is the next higher boiling fraction of crude oil after kerosene, was injected into the gas via a carburettor. The carburettor worked in the same way as that of a petrol engine by spraying a fine jet of hydrocarbon into a gas stream, thereby vaporising and mixing it with the gas. The product was known as carburetted water gas.

Then at the end of the nineteenth century the internal combustion engine was invented. Petrol engines were fuelled by gasoline, a lighter fraction than kerosene, and diesel engines by diesel oil, which is the same as gas oil. Vehicles were (and are) fuelled by light diesel, or light gas oil, and marine engines by heavy diesel, or heavy gas oil.

The term 'fraction' comes from the fractional distillation process, which is the basic petroleum refinery operation. Distillation is carried out in a tower in which the vapour flow is from bottom to top, and the

Table 5.3 Approximate boiling ranges

Fraction or product	Boiling range (°C)
Gasoline	65–90
Benzine	90–140
Naphtha	140–165
Kerosene	165–240
Light gas oil (light diesel)	240–320
Heavy gas oil (heavy diesel)	320–365
Residue	over 365

temperature is highest at the bottom (Fig. 5.27 and 5.28). The tower contains trays on which products that condense over a specific temperature range collect, and these are drained off, usually for further treatment. Table 5.3 shows approximate boiling ranges for the different fractions available from a typical crude oil. It must be remembered, however, that the composition of crude oil varies greatly from place to place, so this tabulation is not universally applicable.

So in the early twentieth century the oil industry changed from a supplier of illuminating oil and gas to a supplier of power. All fractions were now used. The residual oil eventually replaced coal as a fuel for steam boilers on ships. As the use of motor cars increased, so did the demand for gasoline, and means were found (thermal and later catalytic cracking) to break down the heavier molecules and increase the gasoline yields. Refineries grew larger, and crude oil was increasingly transported (in the USA) by pipeline.

Pipelines were an early example of the use of welding for pressurised equipment. The individual pipe lengths were made by forming plate to shape and then making a longitudinal weld in a very large flash butt-welding machine. After removing the flash from the weld the pipes were stretched by means of a hydraulically operated internal mandrel. This operation tested the weld and at the same time increased the yield strength. Then in the field, the pipes were welded end to end by a manual technique known as stovepipe welding, using electrodes that consisted of wire wrapped with paper sealed with sodium silicate. The field welds were porous but this problem was overcome by wrapping with oil-soaked ropes or by encasing in concrete. The pipelining technique so developed has lasted almost unchanged to the present day, except that the field welds are no longer porous.

Otherwise, refineries were of riveted construction. A big change took place, however, after World War II. Fusion welding was introduced as a means of fabricating most items of refinery equipment, together with the

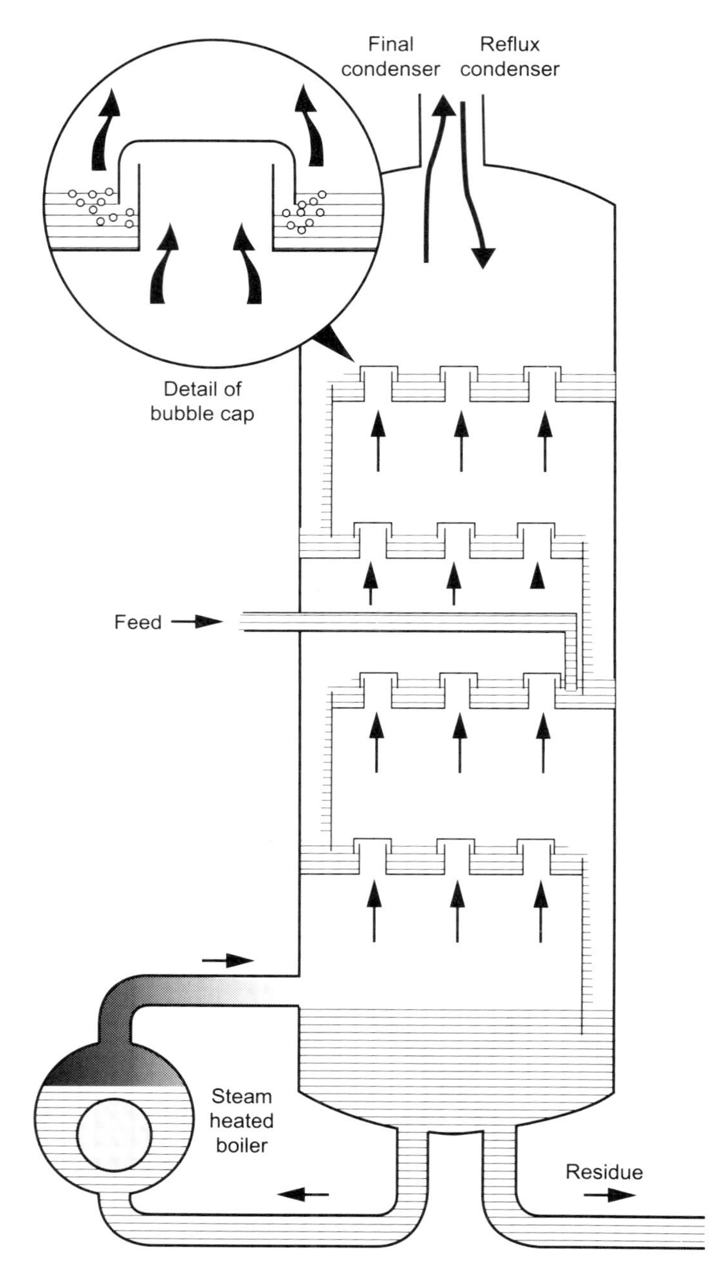

5.27 Distillation (bubble) tower: vapour rising through the tower bubbles through the liquid held on successive trays, stripping out the lighter fractions which are taken towards the top.

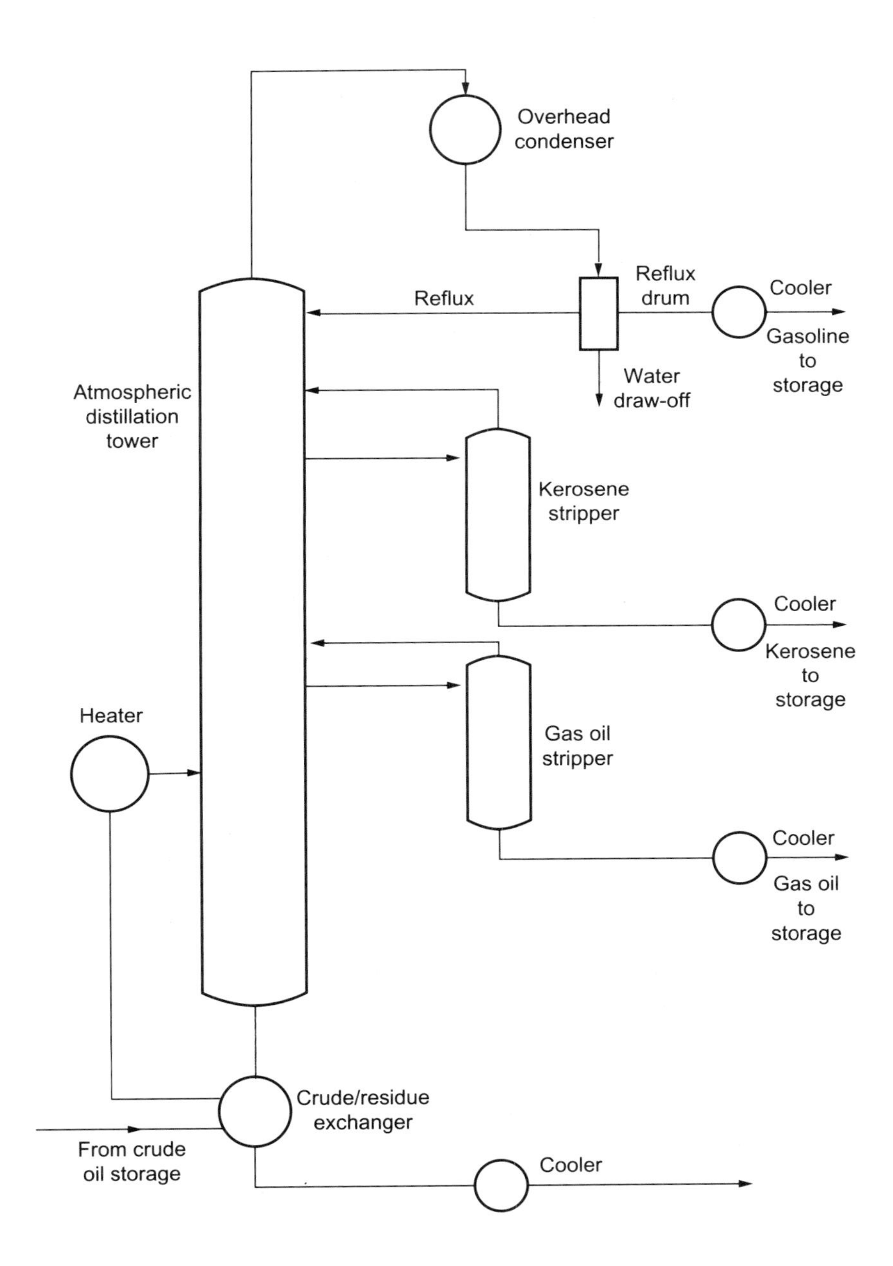

5.28 Crude oil distillation.

interconnecting pipework. The oil industry played a major part in improving welding methods and inspection techniques, and wrote codes for the construction of pressure vessels and pipework. As a result, it became possible to use fabricated process plants for high pressure operations on a large scale.

High pressure technology had been developed early in the twentieth century primarily as a result of the invention of the Haber Bosch process for synthesising ammonia. In this process, a mixture of hydrogen and nitrogen was passed over an iron catalyst at a pressure of up to 200 atmospheres. In order to contain such pressure, it was necessary to use large forgings, and although forged ammonia converters of up to 300 tons in weight were eventually produced, this requirement was a serious limitation.

In the post-1945 years, the techniques that had been developed for hydrocarbon processing were applied to a modified form of the Haber process which operated at a somewhat lower pressure. The final result was a single-train plant using a centrifugal compressor capable of producing 1000 tons or more of ammonia per day, which was much higher than was possible with the old Haber Bosch process. The new process also provided a route to the production of nitric acid. Thus, hydrocarbon processing technology has now provided the most economic means of fixing nitrogen, which was the traditional activity of the heavy inorganic chemical industry.

Hydrocracking is another technique requiring high pressure. This process is used to break down the heaviest of crude oil fractions, including residual oils, to form lighter and potentially more marketable products. The hydrogenation is carried out at a pressure of about 2000 psi in stainless-clad heavy-wall reactors.

One of the major developments of post-1945 years has been the growth of the plastics industry. One of the raw materials for this industry is ethylene which is produced by the pyrolysis of methane at a relatively low pressure. Low density polythene (of the sort used for polythene bags, for example) is produced at high pressure in a reactor made of forged alloy steel.

Alongside these process changes, the technique of fabricating process plants has improved steadily; steels are less subject to embrittlement, and welding techniques are more reliable. On the whole, however, it seems unlikely that such improvements have had much effect on plant safety one way or another.

Losses in hydrocarbon processing plants

The increasing property loss in process plant during the last quarter century or so has been discussed in Chapter 1, where it was concluded that two factors probably contributed to this tendency, namely the increased amount of gaseous hydrocarbon and hydrogen being handled, and increased pressure. To these must be added another factor: the increased production of ethylene, which, as seen in the previous chapter, is relatively easy to detonate.

These greater hazards have not been matched by any radical improvements in the process plant itself. There have, of course, been favourable changes; materials are of better quality, welding is more reliable and the widespread use of quality assurance systems should have reduced the risk of harmful defects in fabricated plant.

These changes have, however, not had the desired result, and as was suggested earlier, it might be better to think about upgrading the standards of design and fabrication of piping handling potentially explosive fluids at elevated pressure. Piping has traditionally been given less serious attention than pressure vessels, on the grounds no doubt that the rupture of a pressure vessel would have more serious consequences than the rupture of a pipe. The records examined here do not support this view; most of the serious damage is due to vapour cloud explosions, and most vapour cloud explosions result from the rupture of piping. The problem at least merits some thought.

Catastrophes: can we prevent them?

The answer to this questions is, of course, no. The evidence shows that in most fields of human activity, and particularly those which are most advanced mechanically, there is a general improvement of safety with the passage of time. Such improvement, however, takes the form of a proportional reduction of accident rate. Thus, if this year's loss rate is 2%, and the annual decrement is 0.03%, then we can expect that next year the loss will be 0.97 $\times$ 2% or 1.94%. So it never becomes zero. Moreover there is always a good deal of scatter in failure rates from year to year, so that the improvement in safety, where it occurs, is not a smooth one.

There is another important factor to bear in mind. In discussing failure rates, we have been concerned entirely with the rate as a proportion of the total number of units in service at the time. This is the only correct way of measuring performance. However, public concern, if it exists, is with the absolute numbers of accidents, on the one hand, and the scale of disasters, on the other, and both these quantities may depend upon factors that are quite independent of the level of safety that has been achieved.

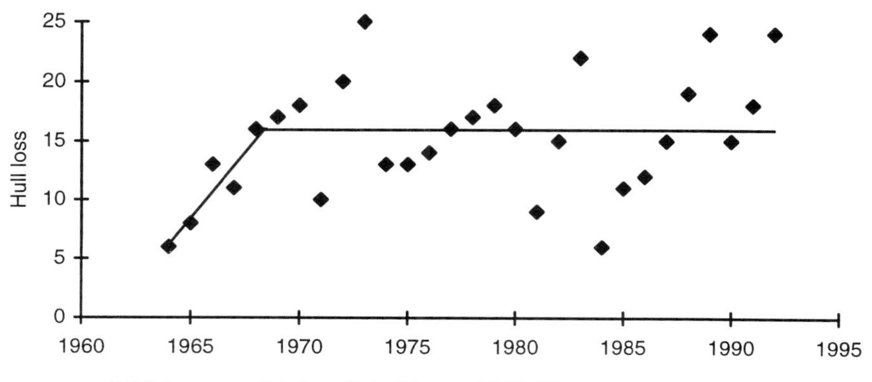

5.29 Passenger jet aircraft: hull losses 1964–92.

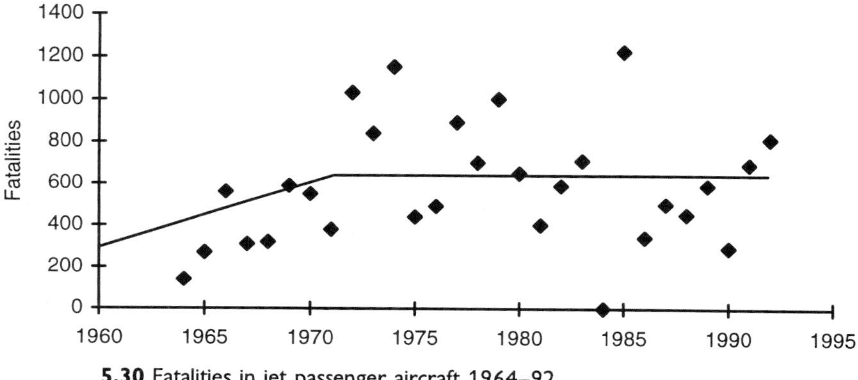

5.30 Fatalities in jet passenger aircraft 1964–92.

Jet aircraft hull losses (that is, total losses) for the period 1964–92 are shown in Fig. 5.29. Initially, there was a rise in the annual loss rate but since 1970 there has been no significant change, although the scatter from year to year is considerable. Fatalities (Fig. 5.30) show a similar pattern, again with a good deal of scatter. Evidently the annual decrement of loss rate is matched by the annual increase in the size of the world fleet of jets.

Shipping losses for the period 1970–92 (Fig. 5.31) follow quite a different course. They increase up to 1980 and then decrease quite sharply. 1980 marked the beginning of a major economic slump in shipping operations, which was paralleled by a slump in oil production. It makes sense, of course, that shipping accidents should increase in numbers during a boom period and decrease during a slump, when the number of shipping movements is reduced. As world trade recovers, it is to be expected that the absolute numbers of shipping losses will start to increase again. There is some evidence that the percentage loss rate is also higher during boom periods and vice versa.

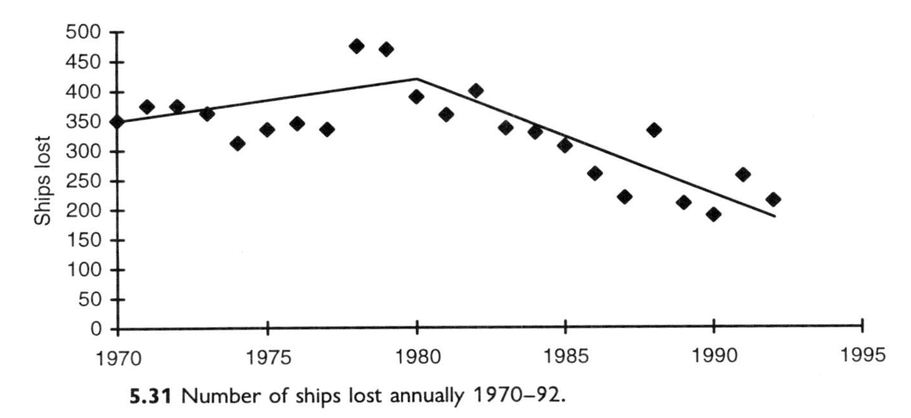

5.31 Number of ships lost annually 1970–92.

The size of catastrophes (other than natural disasters) is very much dependent on the size of individual operating units. The *Titanic* losses were large because, amongst other things, she was a very large ship. Likewise the Piper Alpha losses were many times those caused by a very similar accident in the Mexican Gulf, primarily because the number of men on board Piper Alpha was much greater. In the near future, however, it seems unlikely that vulnerable structures will increase in size to any significant extent. The large passenger ships probably reached their peak with *Queen Mary* and *Queen Elizabeth*, launched in the 1930s. Oil tankers, bulkers, jet aircraft and offshore platforms would likewise appear to have reached the point where further radical growth is improbable. Catastrophes that result from the failure of human artefacts are not therefore likely to increase significantly in magnitude.

Catastrophes that result from natural forces such as flood and tempest do, on the other hand, appear to be increasing, both in terms of the number of people affected and in cost. Most of this increase occurs in the less industrially developed parts of the world. In Western Europe, North America and Japan techniques for flood control, and the construction of earthquake-resistant buildings have made great progress during the last half-century, but it may be a long time before such methods can be applied worldwide.

It is most unwise to be complacent about the integrity of structures. Nevertheless, the records show a decline in the rate of loss, and incidence of human casualties, for most fields of engineering, at least in the more industrialised countries. This benefit does not derive from any single cause or activity; rather, it is an evolutionary trend based on the efforts of countless individuals towards making a better product, be it a skyscraper or a mousetrap, and this trend will undoubtedly continue. The ship of humanity is not, after all, about to founder.

References

1 Wallner F, 'The LD process', *Metal Construction* 1986 **8** 28–33.
2 Shute N, *Slide Rule*, Heinemann, London, 1972.
3 Masefield P G, *To Ride the Storm*, William Kimber, London, 1982.
4 Middleton D, *Civil Aviation, a Design History*, Ian Allan, Shepperton, 1986.
5 Jubb J, 'Structural failure of bulk carriers', Thomas Lowe Gray Lecture 1995, Institute of Mechanical Engineers, London.
6 Lowson M H (Ed), *Our Industry – Petroleum*, British Petroleum, London, 1970.

Appendix: Units

Units of measurement have evolved rapidly during the second half of the twentieth century, culminating in the general adoption (in science and science-based technology at least) of the Systeme Internationale. This system is based on three fundamental units, the metre, the kilogram and the second. A great advantage of the system is its universality; multiplying the unit for force, the Newton, by the unit for length, the metre, gives the unit of energy, the Joule, and the same unit is used regardless of how the energy is manifest. In physics and related subjects this advantage is overwhelming but in other areas of human endeavour it is less so. For the horticulturist for example, there is no particular advantage in designating a 6 inch pot as having a diameter of 150 millimetres. Because of such considerations, possibly allied to innate conservatism, the Anglo-Saxon countries have been slow to adopt the new system.

A particular merit of SI is that it has a separate unit for force, the Newton. In earlier practice a ton for example could mean either the quantity of a substance (as in the displacement of a ship) or it could mean force (in a pressure or stress of tons per square inch). This distinction has been recognised in recent years by the use of the term ton, on the one hand, and ton-force on the other. A ton-force is the force exerted by gravity on a mass of one ton.

The table below gives conversion factors for some customary British and US units of measurements, including those recorded in this book. A more comprehensive list is to be found in the CRC *Handbook of Chemistry and Physics*, published by the CRC Press, Florida. Where appropriate here, and in the text, figures are given in scientific notations, where for example $1 \times 10^3 = 1000$, $1 \times 10^6 = 100\,000$ and so forth.

Abbreviations

J = Joule = Newton-metre
k = kilo = thousand
kg = kilogram

km = Kilometre
ksi = thousands of pounds per square inch
M = Mega = million
m = metre
mm = millimetre
N = Newton = kilogram-metre/second^2
W = watt = Joules/second

Conversion factors

To convert B to A multiply by	A	B	To convert A to B multiply by
9.8692×10^{-6}	atmosphere (atm)	N m^{-2}	$1.013\ 25 \times 10^5$
1×10^{-5}	bar	N m^{-2}	1×10^5
35.3147	cubic foot	m^3	0.0283168
6.1024×10^4	cubic inch	m^3	$1.638\ 71 \times 10^{-5}$
3.280 84	foot (ft)	m	0.3048
23.7304	foot poundal	J	0.042 14
0.737 56	foot pound force	J	1.355 82
$2.199\ 69 \times 10^2$	gallon (UK)	m^3	$4.546\ 09 \times 10^{-3}$
$2.641\ 72 \times 10^2$	gallon (US)	m^3	$3.785\ 41 \times 10^{-3}$
$1.543\ 24 \times 10^4$	grain	kg	6.4799×10^{-5}
$1.341\ 02 \times 10^{-3}$	horsepower	W	745.7
39.3701	inch	m	2.54×10^{-2}
2.362 21	inch/min	mm s^{-1}	0.423 33
0.101 972	kilogram force	N	9.806 65
$1.019\ 72 \times 10^{-5}$	kilogram force/cm^2	N m^{-2}	$9.806\ 65 \times 10^4$
0.101 972	kilogram force/mm^2	MN m^{-2}	9.806 65
3.224 62	kilogram force/mm$^{3/2}$	MN m$^{-3/2}$	0.310 11
0.145 04	ksi (thousand pounds force/square inch)	MN m^{-2}	6.894 76
0.910 05	ksi$\sqrt{}$inch	MN m$^{-3/2}$	1.098 84
		N mm$^{-3/2}$	34.7498
$9.999\ 72 \times 10^2$	litre	m^3	$1.000\ 028 \times 10^{-3}$
0.039 370 1	mil (thou)	μm (micron)	25.4
0.621 371	mile	km	1.609 344
0.031 622 78	MN/m$^{3/2}$	N mm$^{-3/2}$	31.622 78
2.204 59	pound (lb)	kg	0.4536
0.204 809	pound force	N	4.448 22
$1.450\ 38 \times 10^{-4}$	pound force/square inch	N m^{-2}	$6.894\ 76 \times 10^3$
$3.612\ 73 \times 10^{-5}$	pound/cubic inch	kg m^{-3}	$2.767\ 99 \times 10^4$
1.55×10^3	square inch	m^2	6.4516×10^{-4}

To convert B to A multiply by	A	B	To convert A to B multiply by
10.7639	square foot	m^2	0.092 903
0.386 100 6	square mile	km^2	2.589 998
$9.842\ 07 \times 10^{-4}$	ton (UK)	kg	$1.016\ 05 \times 10^3$
$1.102\ 31 \times 10^{-3}$	ton (short ton, US)	kg	907.185
0.100 361	UK ton force	kN	9.964 02
0.064 749	UK ton force/square inch	$MN\ m^{-2}$	15.4443
0.40627	(UK ton force/square inch) $\sqrt{}$inch	$MN/m^{3/2}$	2.4614
0.16506	(UK ton force per square inch)^2 inch*	$(MN/m^2)^2m$	6.0585

*In the text (Chapter 2) this unit is written as 'inch/ton (square inch)^2'. The figure of 250 inch (tons/square inch)^2 quoted for the fracture toughness of the material of the Comet aircraft fuselage is equivalent to 39.2 $MN/m^{3/2}$.

Index